CW01633210

INCUBATION
A GUIDE TO HATCHING & REARING

This book is dedicated to all my grandchildren for helping Nain to feed the chickens.

KATIE THEAR

BROAD LEYS
www.broadleyspublications.co.uk

INCUBATION
A GUIDE TO
HATCHING & REARING

Published by Broad Leys Publications Limited

A catalogue record for this book is available from the British Library.

ISBN: 978-1-910632-30-7

Fourth Edition

Front cover photograph: "Thanamat Somwan/www.shutterstock.com"
Back cover photographs: Eggs - "Cameilia/www.shutterstock.com",
Incubator - Brinsea Products Ltd,
Chicks and Ducklings - Nigel Bowerbank

Unless indicated or credited otherwise, the photographs and illustrations in the book are the author's.

Graphic Design by roger@rogerb.co.uk

Permissions & Advertising
For information about reproducing articles or content from this publication, or to explore the possibility of advertising in or contributing to future editions and other titles by Broad Leys Publications, please visit www.broadleyspublications.co.uk/content

BROAD LEYS
www.broadleyspublications.co.uk

Contents

Maran chick newly emerged from the shell. The downy feathers are still damp. The other eggs are still at the 'pipping' or break-through stage. *Brian Hale.*

Introduction

"That is the black hen's voice", said Mrs Margery. "I'll go and look for her egg".
(The Chicken Market - British Fairy Tale).

Incubation is a miraculous process and the history of artificial incubation is equally fascinating, from the mud brick ovens of ancient China to the sophisticated electronics of today's incubators and hatchers. The new edition of this book not only reflects the main historical developments, but also provides information on the essential aspects of incubation today. There is new material on the selection and assessment of breeding stock, for there is currently concern about the need to preserve utility strains of traditional pure breeds. This can only be done by choosing the best breeders and keeping careful records. Coverage is also given to exotic species of birds such as ostriches, emus and rheas, which are increasingly being kept as farm livestock.

The first part examines the process of development within the egg, and the conditions necessary for successful hatching. The second section of the book covers the brooding requirements of young birds, while part 3 looks at the specific requirements of different species. It is applicable to a wide range of poultry, waterfowl, game and other birds. Part 4 addresses the inevitable question *What went wrong?* Finally, there is a comprehensive reference section.

While providing general and technical information, the approach is also one of practical emphasis. I have attempted to produce a book that is straightforward and readable, in a field where technical jargon sometimes dominates. In this respect, it is suitable for schools as well as commercial and amateur breeders.

I had no idea, when I wrote the first edition of this small book in 1987, that it would achieve sales all over the world, including areas where there is no mains electricity available, and where 'making do' is part of the culture. I am most grateful to all those who wrote to me about their experiences, or who sent in suggestions. I am also indebted to the contributors of *Country Garden & Smallholding* (previously *Home Farm*) magazine who have contributed their ideas, experiences and DIY tips.

Thanks to the help and cooperation of manufacturers, there is also a survey of incubators, hatchers, brooders and other equipment in this edition.

I hope that the book can justifiably claim to be a useful reference work for breeders large and small, presented in a readable form.

Katie Thear, Newport, 1997.

A cabinet incubator with automatic egg turning and adjustable turning periods, automatic temperature and heater controls, and a 'fail-safe' back-up heating system. The tray at the bottom is a separate hatcher to which eggs near the point of hatching can be transferred. *Curfew Incubators.*

History of incubation

The temperature for incubation is checked by placing the egg against the eye.
(Leslie Pearce-Gerris, 1938)

There is nothing new about artificial incubation. Four thousand years ago the ancient Egyptians were familiar with the concept, using large oven-type structures made of mud bricks. They were not heated by burning camel dung at this time as many books have stated, because the camel was only introduced to Egypt at the time of the Arab invasions, about 1,000 years ago. What is true to say, is that camel dung appears to have become the main fuel source after the Arabs took over Egypt, and the same incubation practice continued well into the present century.

Leslie Pearce-Gerris, writing in 1938, described a visit to an Egyptian hatchery, called a *mahmil*, after overcoming the workers' *'superstitious fear of visitors who might cast an evil eye on the hatching eggs'*.

'A vaulted and domed passage runs the full length of the building, in the roof of which are holes about the size of dinner plates. These are for ventilation, but one is given the impression of being in a church crypt. There are on either side of this passage, the incubators. They are part of the building construction, and each is unconnected with its neighbour - with two vaulted doors immediately above one another. If the lower door were opened, it would be found that here the eggs were stacked in the early stages of incubation, and the floor of one of these compartments will hold well over five thousand eggs. Four fires were kept burning, one in each earner of the upper chamber, the fuel used being rice or bean straw and camel dung. The temperature for incubation is checked by placing the egg against the eye.'

The same families had apparently been in charge of the hatcheries for many generations. The accumulated experience passed on from father to son enabled an extraordinary degree of expertise to be shown, bearing in mind that there was no thermostatic control other than the placing of an egg on the eyelid to determine its temperature.

In China, charcoal heated mud brick ovens were also used, but eggs were placed in large wicker baskets which were then lowered into the heated areas. This was probably a development of an earlier technique of placing eggs in hot beds of manure, a practice still occasionally to be found in parts of Malaya. It works on the simple principle that the lower down in the heap the eggs are placed, the higher the temperature. Sand was placed over the manure so that the eggs did not become contaminated. It may have been the snake laying its eggs in similar situations which first provided the idea thousands of years ago.

Fig 1. Traditional Still Air Incubator - heated by paraffin

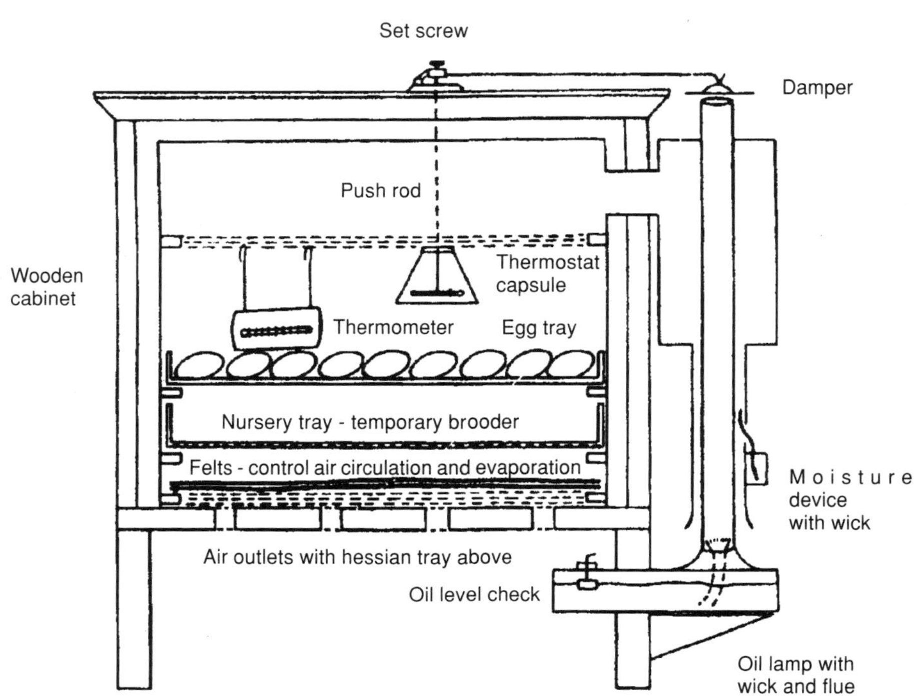

Fig 2. Thatched roofed,
underground incubator house
widely used in Britain and Europe
until World War 2

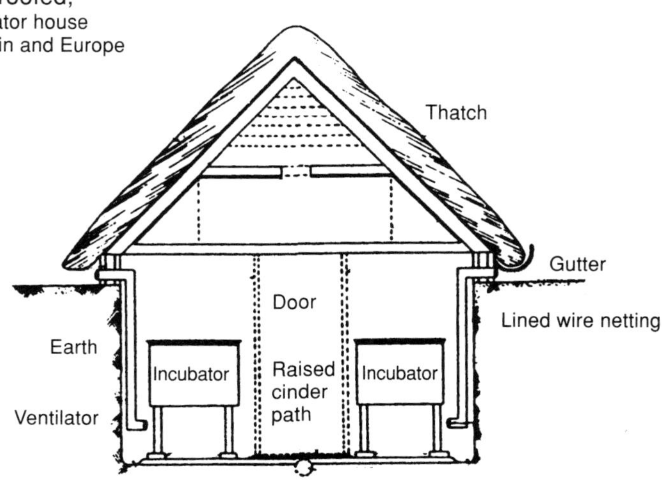

Europe does not appear to have awoken to the possibilities of artificial incubation until the eighteenth century when rotting horse manure was used as a heating agent. Variations on the principle of a box subjected to heat were developed, with hot air and hot water being used, but there was no simple method of controlling the temperature, and manual checking was necessary every few hours.

In 1851 an incubating machine was shown at the International Exhibition in England. It was basically a box with holes in the bottom, with a thick bed of hay. Eggs were placed in the hay and covered with a thick layer of feathers. Finally, a rubber hot water bottle was placed on top. A thermometer was used to check the temperature regularly, and the hot water bottle needed constant refilling.

It was in 1883 that the break-through came with Charles Hearson's invention of the capsule control, the first thermostat. This was a hermetically sealed metal envelope containing ether and alcohol. This remained normal at a temperature of 104°F. When subjected to heat, it expanded while a lower temperature caused contraction. By connecting the capsule to a system of rod and levers going to the lid covering a lamp flue, Hearson made it possible for the lid to be raised when the temperature became too high and to be lowered when the temperature dropped. With the former, hot air was dispersed into the room. With the latter, the hot air was forced to pass into the chamber heating the water.

Paraffin (kerosene) was widely used as a heat source until comparatively recently. (At time of writing there is still one manufacturer in the UK who is producing a paraffin model, as well as a range of spares). However, electricity is making these machines redundant, except in remote areas of the world where paraffin and battery-operated incubators are of great significance.

Different materials have been used over the years, with varying insulation properties. Metal and wood were common, and thatched roofs were often used in commercial hatcheries early this century. These were often lined with asbestos which was probably a bit of an 'over-kill' in more sense than one. In recent years the availability of plastics and wood composites have made small incubators energy-efficient and easy to clean.

Thermostats were to improve over the years, particularly with the invention of the bi-metal strip which worked by having two different metals conducting heat at different rates, thus causing uneven expansion of the two sides. This type of thermostat was ideal for electric powered incubators and is still in use in many machines today. With the advent of the electronic thermostat, however, thermostatic control has achieved a level of accuracy previously undreamt of.

Incubators have used all manner of power; camel and horse dung, charcoal and paraffin, gas and electricity. They have come a long way in the last few thousand years, but no-one would claim that an incubator has been invented which could equal that most efficient and successful machine - the broody hen!

Fig 3. Female reproductive system

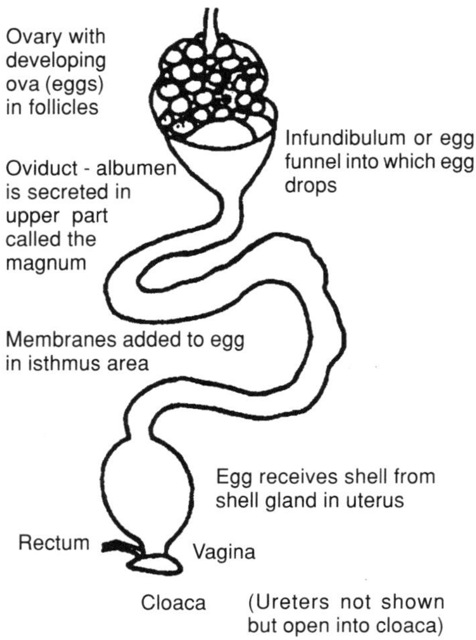

Fig. 4 Male reproductive system

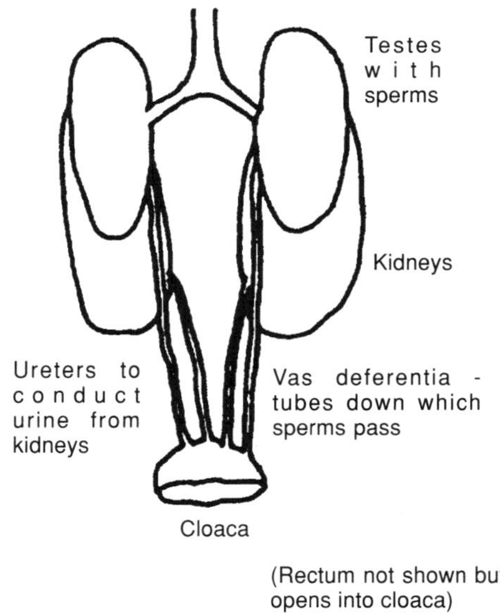

Fig 5. Structure of the egg

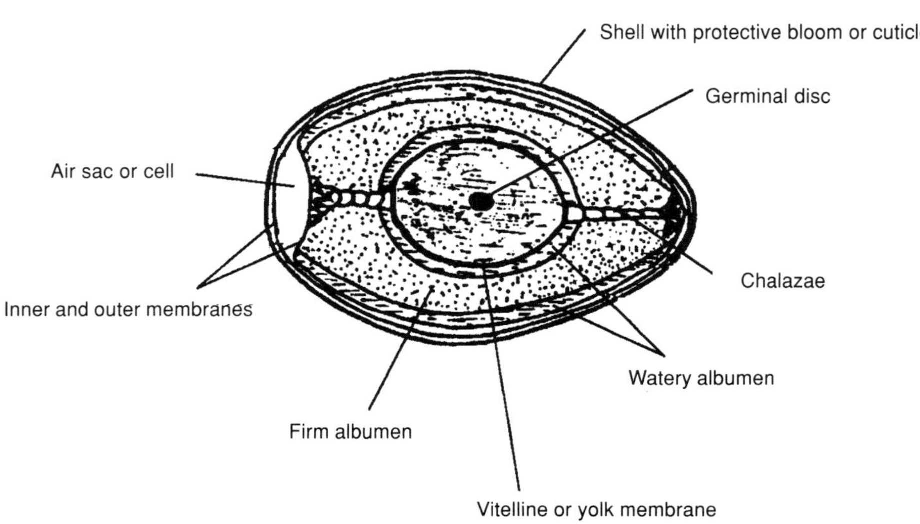

Part 1
Incubation and hatching

"A hen is only an egg's way of making another egg".
(Samuel Butler)

The egg

An egg is a miracle of creation; a genetic blueprint within its own protective walls, equipped with the means of survival and transformation into a complete organism in a matter of weeks. Eggs come in different sizes, depending on whether the source is a quail, a chicken or an ostrich, but the basic structure is the same. There are essentially six parts to the egg: the shell, shell membranes, albumen (white), yolk (yellow), chalazae (supporting tissue) and germinal disc (fertile area).

Shell

This is composed mainly of calcium carbonate in a rigid, irregular ovoid shape. One end is normally more pointed than the other so that when the egg rolls, it does so in a circular manner, ensuring that it does not roll too far from its original position. In the wild, this is a useful feature, contributing to the survival of eggs which might be laid on precipitous ledges. It is not a feature found in all eggs, of course. An ostrich egg is relatively rounded at both ends, and so, strangely enough is that of the Pekin duck!

The shell is porous, allowing the interchange of oxygen, moisture and carbon dioxide. The incoming air provides oxygen necessary for respiration, while carbon dioxide, as a waste product diffuses back to the air through the pores. These same pores also control water exchange, allowing excess water vapour produced during respiration to escape, yet allowing moisture in the air to enter from the outside when necessary.

The shell is a highly effective protective wall, denser on the outside than the inside so that the emerging chick finds it easier to break out than it would be to pierce from the outside. The inner layer also provides the calcium source for the developing bones of the chick. At the time of laying, the egg is protected by a thin, non-cellular coating sometimes referred to as the 'bloom'.

Yolk

This is essentially a food store made up of water, protein, fats, vitamins and minerals, contained within a vitelline membrane. Contrary to what is popularly believed, comparatively little of it is used up during the incubation period. Most of it acts as a reserve for the first few days of life of the hatched bird. It is absorbed into the

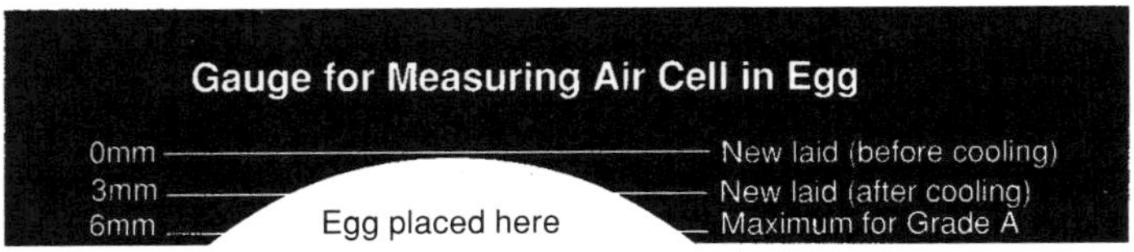

The gauge is held over the blunt end of the egg while shining a light through it. The relative height of the air cell can be seen by comparing it with the scale.

Fig. 6. Measuring the height of the air cell to determine relative freshness

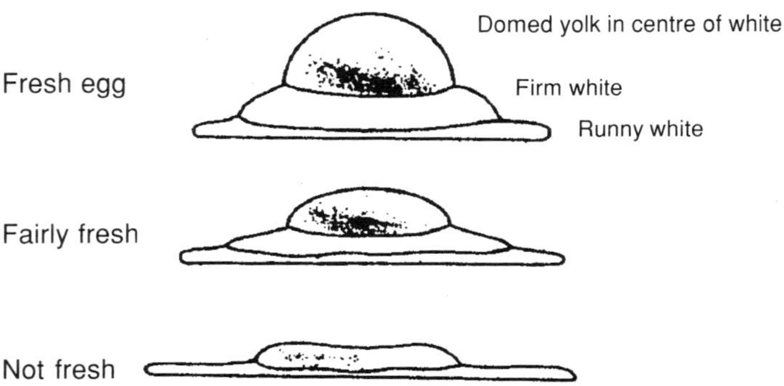

Fig. 7. Egg structure as an indication of relative freshness

abdomen of the chick just before hatching. This explains why day-old chicks can be transported so successfully: they can survive quite adequately without food for a couple of days.

Chalazae

These are spirals of supporting tissue within the albumen, connected at each end of the egg. They hold the yolk in its central rotating position.

The threads are coiled in opposite directions so it is important when turning the eggs during incubation, that they are not always turned in the same direction, otherwise the effect is to coil one chalaza more tightly, while the other is gradually uncoiled. The eventual result would be to disrupt the ability of the yolk to rotate freely. It is also important to rotate the egg slowly and with great care (as for example, when checking its contents during incubation) so that no damage is done to the developing embryo.

Shell membranes

There are two of these, separated at the broad end by an air sac or air cell. The external one is attached to the shell, while the internal membrane, although loosely connected to the outside one, can separate at the broad end of the egg. This allows the air space to increase as other cell contents are used up by the chick, ensuring an adequate supply of air for the chick while it is still in the shell.

When an egg is first laid the air cell is barely discernible, but as it cools the inner cell membrane contracts and pulls away from the outer one, leaving the air cell space in between. As the egg gets older, the air space increases, regardless of whether it is a fertile egg or not, because it is gradually losing moisture.

If hen's eggs are being sold and described as 'class A', the height of the air cell must not exceed 6mm when viewed against a light. (In the USA the designation is 'AA' with a maximum of $^1/_8$"). If it is more, it is an indication that the eggs are not fresh. (See *Fig. 6*).

Albumen

The albumen or 'white' of the egg is composed of protein, water, vitamins and minerals. This food store is used up by the developing embryo during incubation.

It is made up of three layers, thin inner and outer areas of watery white, and a central more jelly-like layer. The watery outer layer allows for an efficient diffusion of gases with the outside. The central, jelly-like albumen provides a protective cushion and acts as a shock absorber. The inner watery layer lets the yolk and germinal disc rotate freely, ensuring ready contact with food and oxygen.

It is easy to see the different whites when a newly laid egg is broken (See *Fig 7*), but as the egg ages and settles, they become indistinguishable.

Germinal disc

This is the fertile part of the egg where the male cell unites with the female one, leading to cell division and embryonic development, as long as conditions are right. It is sometimes called the egg cell.

Formation of the egg

A hen's ovaries are situated at the back of the adbominal cavity. They look rather like clusters of grapes of differing sizes (See *Fig. 3*). Each 'grape' is a yolk with an unfertilized *ovum*, and the larger it is, the more ripe it is. As each egg ripens, it moves into the *infundibulum* - the funnel-like top of the oviduct passage which eventually leads to the *uterus*.

It is in the upper *oviduct*, just below the infundibulum, that fertilization takes place. This follows mating where the male sperm travels up the oviduct and fuses with the female cell on the yolk. It has been estimated that sperms will remain alive for 2-3 weeks after copulation, waiting to fertilize more ova as they descend.

The fertilized egg moves down the oviduct, acquiring a covering of albumen in the *magnum* area. In the *isthmus* region of the oviduct it is coated with the shell membranes. When it reaches the uterus, the hard shell is secreted around it, and the egg moves down to the vagina to be expelled at the cloaca. Taking into account different species and relative size, the process is essentially the same in all birds.

The formation and passage of an egg, from ovary to cloaca, generally takes between 24-36 hours, but there are variations depending on species and environmental conditions. If it has been fertilized, cell division begins almost immediately, so it is important to store the egg in cool conditions until it can be incubated, otherwise there will be partial and disrupted development.

When first laid, the egg is damp and any surface markings or pigmentation can be rubbed off. If you don't believe it, try it for yourself! If you have Maran hens or similar, which lay pigmented eggs, see if you can be present just at the point of laying. Take the egg *while it is still damp* and rub with your thumb. The markings will be rubbed off! Sometimes even the movement of the hen causing the egg to rub against nesting material will rub off the markings. Once the egg has dried, however, no amount of rubbing, or even boiling, will have any effect. The drying of the egg surface provides a protective 'bloom' which helps to keep out dirt and infection.

Egg collection

Eggs that are to be incubated should be collected as frequently as possible in order to minimise the possibility of their getting dirty. Providing clean nest boxes and nesting material is also important in helping to ensure that eggs are kept as clean as possible. Wood shavings or sawdust from untreated wood work well, as long as they are changed frequently, and that every effort is made to exclude droppings. This may mean having to shut off the nest boxes at night if there is any indication that some birds are using them to sleep in at night.

Use a basket or rubber bucket to collect the eggs, for it is important that they should not be jarred or cracked. Some clean kitchen roll, or wood shavings in the case of the bucket, can be used as a liner for added security.

It makes sense to wash your hands before collecting eggs, using an anti-bacterial spray, so that the chances of introducing pathogens are minimised. Close-fitting rubber, surgical gloves are often used by breeders for the same reason. Whenever the eggs are subsequently handled, whether it is for manual turning or candling, the same degree of care is required.

Cleaning

If the eggs are dirty, they can be brushed clean with a dry nail brush or sandpaper. Particularly stubborn marks may need to be scraped off with a scalpel. This form of cleaning deals with visible contamination, but what about microscopic pathogens that can rapidly multiply and cause disease in the incubator?

Eggs should be washed in water to which an egg sanitant has been added. There are several brands available from incubator suppliers, and it is important to follow the instructions for the particular product as they can vary in strength. It is also vital to use warm water so that any bacteria on the egg are drawn away from the

Fig 8. Washing eggs that are to be incubated. *Patrick Pinker Game Farm.*

pores. If the water is colder than the egg, the effect can be to draw *in* bacteria! If you have a lot of eggs to clean at a time, remember to replace the water regularly, in case it gets too cold. Leave the eggs to drain and dry in plastic inserts. These are manufactured to fit into incubator trays and are available in a range of egg sizes. They are also useful if you have only a small number of eggs in a manually-turned incubator, for they stop the eggs rolling about - and possibly taking up the wrong position when you have just turned them. Most incubator suppliers sell them.

Storage

If there are not enough eggs to fill the incubator, the eggs will need to be stored until there are sufficient numbers. The egg inserts referred to earlier can be used to hold them. Alternatively, normal, clean egg cartons or Keyes trays (known as flats in the USA) are suitable.

The eggs should be placed, broad end upwards, in a cool, draught-free room within a temperature range of 15°-18°C if they are to be incubated within a week. If they are to be stored longer, adjust the temperature to the 12°-15°C range. Tilt the carton or tray daily, ensuring that it is tilted in the opposite direction to what it was previously so that the embryo does not adhere to one side.

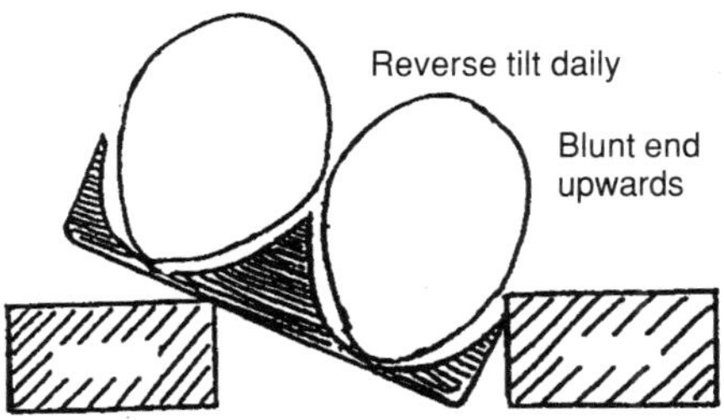

Fig 9. Storing eggs prior to incubation

Eggs should be incubated as soon as possible after laying, ideally no later than a week. Eggs that are older than this may still hatch but hatchability declines after the first two days, at a rate of 2% per day.

Before the eggs are incubated, it is a good idea to dip them in an egg sanitant before placing them in the incubator, as an added precaution against disease-causing pathogens. Allow them to come up to normal room temperature for 24 hours before incubation is started. This prevents any abrupt change of environment which may cause 'shock' to the eggs.

Selection

So far we have assumed that all the eggs are destined for the incubator, but this should certainly not be the case if a good rate of hatching is required. There needs to be a process of selection with clear criteria in mind of what constitutes a good egg. These criteria include size and shape, colour and quality of egg shell, and whether the eggs have come from good breeders.

No cracks

Eggs for incubation should be undamaged and free of hairline cracks. Some of these can only be detected by candling, or holding the egg up to a bright light, so it is a good idea to do this during the process of selection. Those that fail the test should be discarded. Having said that, it could be that eggs of particularly rare breeds are so valuable that even cracked eggs are given a try.

I have successfully incubated cracked eggs of Maran chickens by the simple expedient of painting a little colourless nail varnish over the crack. If you try this, avoid getting the varnish over any other part of the shell. Another technique which has been used successfully by a breeder I know, is to put a tiny strip of ultra-thin sellotape over the crack.

Incubating cracked eggs is, of course risky, for there is a greater chance of disease-causing bacteria having entered. It is a procedure that should only be tried with a small number of eggs, rather than subject a large batch to the risk of disease.

Size and weight

Generally, the best size of egg for incubation is one which falls into the medium range for its type. For the average chicken egg, this would be in the 53g-63g band. Evidence indicates that over-large or particularly small eggs can lead to problems. Large eggs are more likely to be double-yolkers which could be problematic in the incubator. This has to be balanced against the fact that you might be selectively breeding *for* layers of large eggs, so it is a matter of judgement. In any case, candling the egg with a bright light will reveal whether it has more than one yolk.

Very small eggs may be devoid of a yolk, a condition known as a 'wind' egg, unless the eggs under discussion happen to be from bantam breeds.

In most cases, the size can be determined by eye, but weighing eggs before they go into the incubator is a good idea because the weight loss during incubation is an indication of the required level of humidity at any given point. Knowing the weight at the start of incubation enables one to work out fairly precisely whether the humidity level is correct. (See the sections on humidity for more details).

Weight loss for most poultry and waterfowl eggs is 11-13% during incubation, while that for ratites is 13-15%.

Shape

The shape of an egg is also important because of the placement of the air cell. If the egg is too rounded, the air sac may be displaced so that the chick is unable to breathe during the vital 'pipping' or breaking out stage. Bear in mind that some eggs, such as those of the ostrich and Pekin duck, are naturally rounded and this will need to be taken into consideration. (The ostrich chick kicks its way out).

Eggs of the same size and shape, and from the same species, should be incubated together, to ensure even hatches. Similarly, each batch should be started at the same time. Avoid the temptation of 'slipping in another few eggs' after starting.

Fig 10. Shell quality characteristics

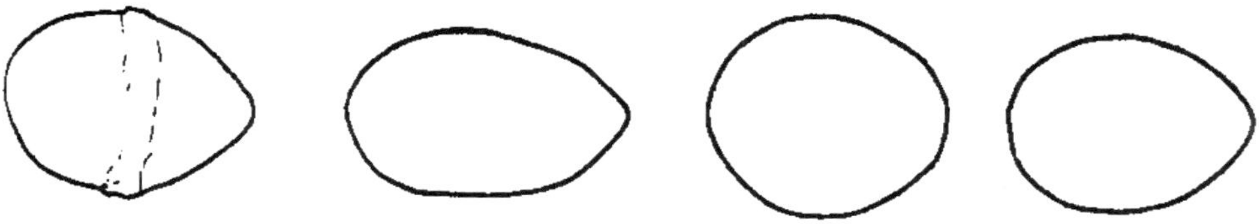

Ridges on shell. Discard.

Overlong with straight sides and possibly weak patches. Discard.

Too round. Air sac likely to be displaced. Discard.

Average size with good shape and colour. Select.

Shell quality

The texture of the shell should be smooth and free from chalkiness, blemishes and ridges. These may indicate poor nutrition in the hen, or some form of stress. A sudden shock, such as loud noise or heavy rain, can disrupt the passage of the egg down the oviduct, delaying it, so that it gains an extra ridge of shell material. Ridges and thin patches may also be because the hen is getting old or has had infections. It indicates how important it is to ensure that only the best, healthy birds are used for breeding purposes. This is an aspect that is covered later.

It is worth mentioning again, that no matter how good the quality of eggs, rough handling is likely to damage them so that hatchability is drastically reduced. Similarly, exposing them to dirty conditions and handling increases the chances of disease, death in shell and poor hatches. The key words are *gentleness, cleanliness* and *care.*

Shell colour

Shell colour is a genetic factor. A Leghorn produces white eggs while a Rhode Island Red lays pinkish brown ones. Some breeds, such as the Maran and Welsummer, produce dark brown, speckled eggs. This like most things, is a generalisation, for there is a considerable variation, depending on the particular 'strain' of bird. A poor example of a Maran may lay eggs that are not much darker than those of a hybrid, and with few speckles, while a good example will produce chocolate brown eggs.

Selection for egg colour will therefore be dependent on the breed of bird, as well as on breeder preference if he or she is carrying out a programme of selective breeding.

Assessing the colour and shine of the shell can be done by eye or with the use of a reflectometer which gives a comparative reading. Dark, patterned eggs are obviously more difficult to candle than light, uniformly coloured ones.

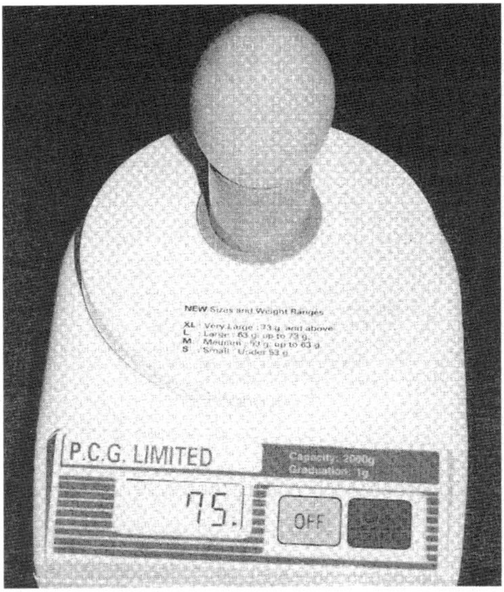

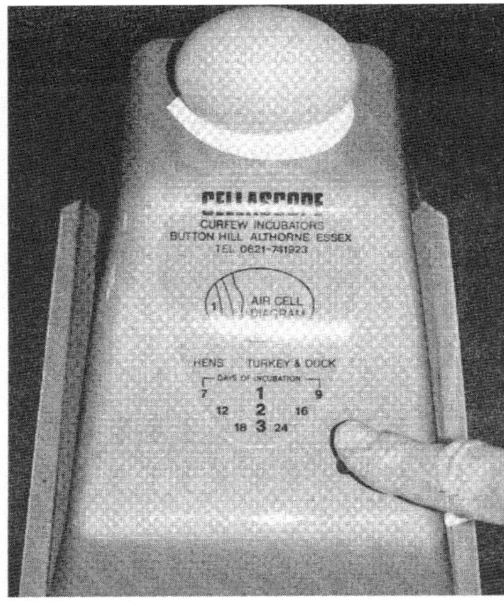

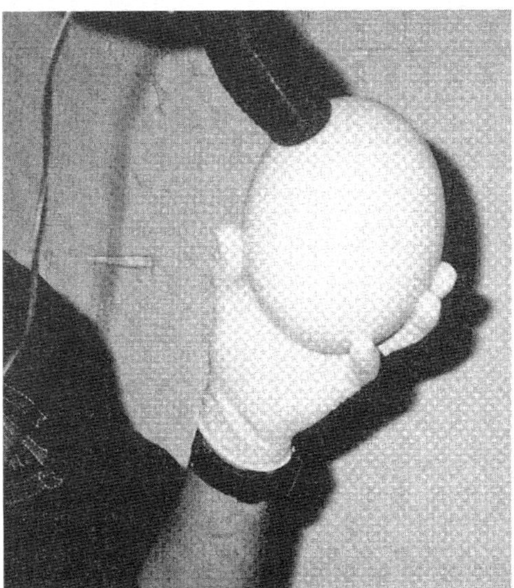

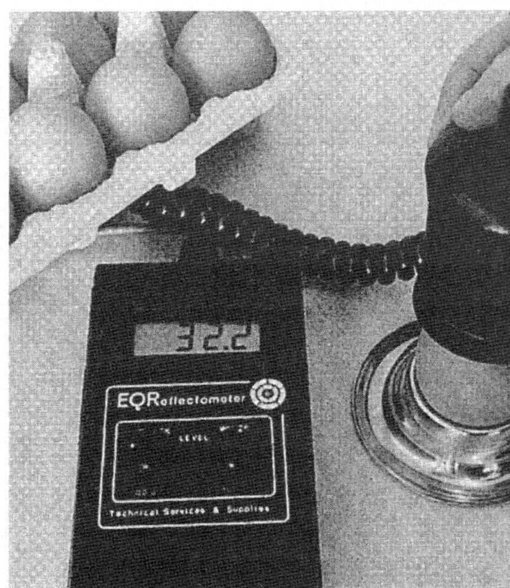

Top left: Part of the egg assessment procedure is to weigh them. This hen's egg weighs 75g and is therefore in the *Very Large* bracket. Normally, medium sized eggs are the best choice for incubation, but in this case, the egg will be selected. This is because the hen concerned is part of a selective breeding programme for the improvement of egg laying strains, and large eggs are therefore an asset.

Top right: Here, the same egg is being candled where a bright light is shone through it to make sure that the contents are normal - with no double yolks - and that there are no hairline cracks in the shell. Later, the candler can be used to keep track of internal development so that any infertile eggs can be removed from the incubator. It is also used to check the size of the air cell at different stages of incubation so that the correct level of humidity can be provided.

Bottom left: Using a hand candler to check an ostrich egg. Note the surgical glove to maintain good hygiene.

Bottom right: Using an *EQR* reflectometer to assess shell colour. A good brown egg should read between 25-35%. *Technical Services and Supplies.*

The breeders

A breeding pen of Vorwerks, an attractive German breed, at *The Wernlas Collection.*

Fertility and hatchability are not the same thing. *Fertility* is the ability to produce viable sex cells - ova or spermatozoa - and to mate sucessfully so that an embryo is formed. *Hatchability* is the relative chance of that egg developing and hatching, for even if the birds are fertile, a defect of any kind, whether genetic or environmental, will reduce the chances of successful hatching. It is also necessary to have a clear understanding of what is required of the progeny, for *show characteristics* are not necessarily the same as *utility features*. The former will be a strict adherence to the *Standards* of the fancy sector, while the latter places more emphasis on the productive capabilities of the birds - are they good layers or table birds? The choice of breeding birds is vital, but it is not an easy task.

The outward appearance of the bird is referred to as its *phenotype*. A Rhode Island Red, for example, can be recognised as such by what it looks like, and whether it conforms to the *Standards* set down for the breed by the breed club.Its *genotype* or genetic makeup cannot be seen, and while it may be an excellent bird for show purposes, it may lack the necessary genes for good production. The only way in which this factor can be determined is by a painstaking process of selection of a particular productive line of birds, and by keeping careful records. It is then possible to say, *"This bird is a good example of the breed's phenotype and will probably be a good layer because it has been bred from a productive strain"*.

It may at some time in the future be possible to carry out a test to determine the DNA profile of an individual bird, so that its *genotype* is as visible as its *phenotype*. For most breeders however, particularly those concerned with the conservation of pure breeds, the approach is still the traditional one of selective breeding.

Assessment of potential breeders

Only the very best birds should be chosen for breeding. The following criteria are important in their assessment:

Health

The male and female chosen for breeding should be sound, healthy and vigorous. Visual checks can also be carried out on any potential breeders, for the external appearance is often an indication of health, as the illustration opposite indicates.

They should also have been tested to ensure that they are not carriers of the particular form of salmonella that can be transmitted via the egg to the progeny. In addition, only those birds that have been vaccinated against Marek's disease when they were chicks should be used as breeding stock. The vet will advise on the appropriate course of action for a particular area, depending on whether there are endemic diseases in certain geographical locations.

If there is any doubt about the status of a potential breeding bird, a veterinary surgeon or veterinary laboratory will arrange for the bird to be blood-tested. Testing for salmonella carriers can be carried out on site, using the equipment supplied by the vet or laboratory.

Within the European Union there is a legal requirement for all those with a breeding flock of 250 birds or more, to test them for salmonella.

Good example of the breed

Breeders should be good examples of their particular breed. Each breed has its own *Standards* drawn up by the breed club. These describe the ideal characteristics to aim for. The breed secretary will provide details of the appropriate standards. In the UK they are also detailed in the publication *British Poultry Standards*. Waterfowl are covered by the *British Waterfowl Standards*. Other countries have their own standards. It should be remembered however, that standards have largely been drawn up for the fancy, and there may be a conflict between this and the utility aspect. If the priority is to produce productive birds, the most important assessment, after health, will be whether the bird is from a good, utility strain. Utility breeders have always stressed that *'strain is more important than breed'*. The ideal situation, perhaps, is to have a good measure of both.

Fig 11. Characteristics of a good healthy breeder

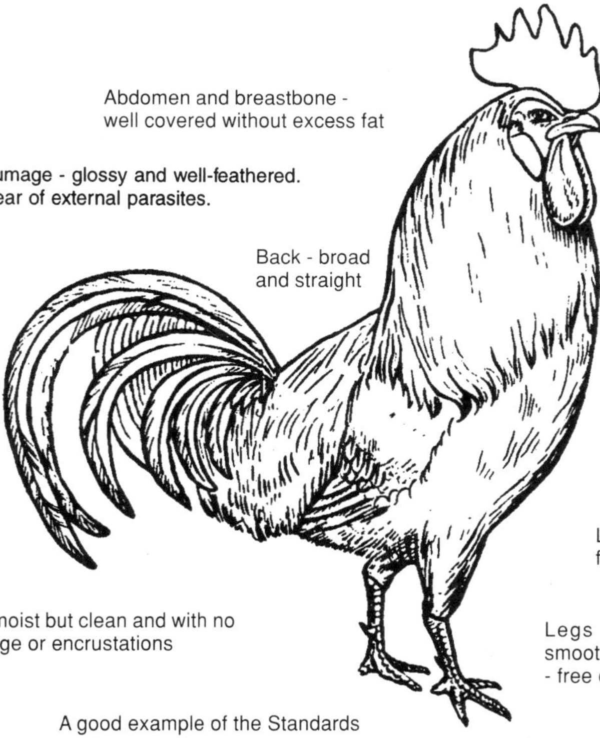

Productive capacity

If hens are to be chosen for their productive capacity as layers, there are certain visual characteristics that can be used as a guide, in addition to the general characteristics for both sexes shown above. The dimensions between the pelvic (pubic) bones, and between the pelvic bones and keel will be more generous in a hen that is in-lay. There are three areas to check - the relative distance between the pelvic bones, that between the base of the tail and the end of the keel (breast bone), and the space between the last rib and the pelvic bones. At least two fingers width should fit between the pelvic bones, while three or more fingers should fit between the pelvic bones and tip of the keel. In the non-laying bird the pelvic bones are close together, while the vent is small and dry.

In yellow-skinned hens it is possible to estimate past egg production from the amount of yellow pigment left in the body. This is normally found in the vent, beak, legs, feet, ear lobes and eye lids. As laying commences, less pigment is deposited in these areas and the colour gradually fades - a process called 'bleaching out'.

Fig 12. Assessing a good layer

The widths shown are relatively generous when compared with a poor layer

Bleaching takes place in the following order: edge of vent, edge of eye lids (called 'eye ringing'), ear lobes, corners of mouth, rest of beak, feet and shanks, hocks. It takes 4-6 weeks for all the pigment to go from the beak and around 4-6 months for it to go from the feet and legs.When laying ceases, the pigment returns in the same order that it left, but at a faster rate. Its presence is an indication of how long a hen has been out-of-lay.

A non-laying hen usually has a comb and wattles that are smaller, paler and more shrivelled than one that is laying.

Good producers tend to delay moulting for a longer period than poor layers, and then be rapid moulters, taking around 8 weeks in all to re-feather. Poor birds often have a slow moult, taking up to 18 weeks to re-feather.

When assessing pullets for productive potential any 'crow-headed' birds with wrinked faces should be discarded for breeding. Those with large, bright combs and bright, prominent, alert eyes are potentially worthwhile

It should be emphasised, however, that in order to develop a good 'strain', selective breeding will need to be carried out over a number of generations. This involves trap-nesting.

Trap-nesting

The eggs of individual females will need to be identified by a system of trap-nesting. A trap-nest is where a door closes behind a bird as she enters the nest and keeps her there so that her egg can be identified. The practice involves identifying individual birds by means

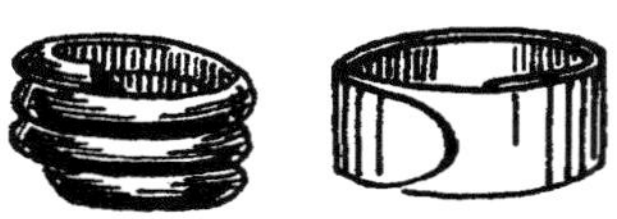

Fig. 13. Poultry leg rings in different colours.There are also numbered ones available.

A trap nest which allows entry into the nestbox, but closes behind the hen and keeps her in the nest until the egg can be identified as hers when she is released. The egg is marked with her unique leg ring number for identification and recording.
Another design of trap-nest is shown below.

of leg rings which are available in an assortment of colours. It is also possible to buy numbered leg rings so that a bird can have a unique number. This is essential for keeping breed records. Trap-nesting does impose a rigid regime. Very frequent checking is necessary so that no bird is kept confined for longer than necessary, on humanitarian grounds.

Make your own trap nest

This is a design for making your own trap-nest by Steven Rogers. It was first published in *Country Garden & Smallholding* magazine.

Cutting list This assumes a trap-nest whose width, depth and height are all 500mm is being made. This size will be suitable for all bantams and most large fowl, but all dimensions can be either reduced or enlarged, depending on the specific breed being 'trapped'.

4mm thick exterior grade plywood

1. Base 500 x 500mm (1 of)
2. Sides 500 x 500mm (2 of)
3. Front 500 x 500mm (1 of)
4. Back 510 x 500mm (1 of)
5. Lid 570 x 570 mm (1 of)
6. Slats 230 x 40mm (6 of)

Screws and Nails, etc.

25mm x 38mm planed treated timber (finishes approximately 20mm x 33mm)

7. Lid edges 550 mm (4 of)
8. Base edges 480mm (4 of)
9. Side uprights 460mm (4 of)
10. 'Feet' 400mm (2 of)
11. Trap door sides 340mm (2 of)
12. Trap door top 240mm (1 of)

Use either 6 x ¾ zinc plated screws or ¾" panel pins and a waterproof wood adhesive.
50mm wide hessian webbing - for holding trap floor slats.
6mm staples or similar - for fixing slats to webbing.
Roofing felt, corrugated *Perspex* or *Onduline* - for lid if the nest box is to be used outside.
Wood preservative - either creosote or a water based wood preserver, if preferred.

Trap nest assembly

1. Fix each of the four base edges to the base, narrow side of timber to plywood. Fix the two 'feet' to the opposite side of the base 50mm in from the edge.

2. 100mm from the top edge of each side, cut out a 100mm x 25mm handle and six 25mm diameter air holes. Fix the two side uprights to each side, wide side of timber to plywood, flush with the top, which will leave a 40mm space at the bottom.

3. Cut a 300mm x 200mm hole centrally in the front, 50mm up from the bottom edge. Fix the two trap door sides, narrow side of timber to plywood, to the inside of the front, flush with the bottom of the hole and 20mm away from the edge. Before fixing the trap door top to the plywood, place two strips of 50mm wide webbing between the trap door top and the plywood. Using adhesive and staples, fix the slats to the webbing, leaving a 10mm gap between each slat.

4. With the base standing on its feet, fix the two sides to it. The side uprights will rest on the base edges and there will be a slight overhang at the bottom. Lay the trap-nest on its front and fix the back to the two sides and the base. There will be a slight lip on the left and right hand edge which can be sanded off when the adhesive is dry. Turn the trap nest over and fix the front in the same way. Stand the trap nest on its feet and check that it is square. Allow to dry before sanding off all rough edges.

5. Fix the lid edges in the same way as the base edges. Fix prepared lid covering, if the trap-nest is going to be used outside.

6. Treat inside and out with wood preservative and allow to dry thoroughly before use.

Trap-nest assembly

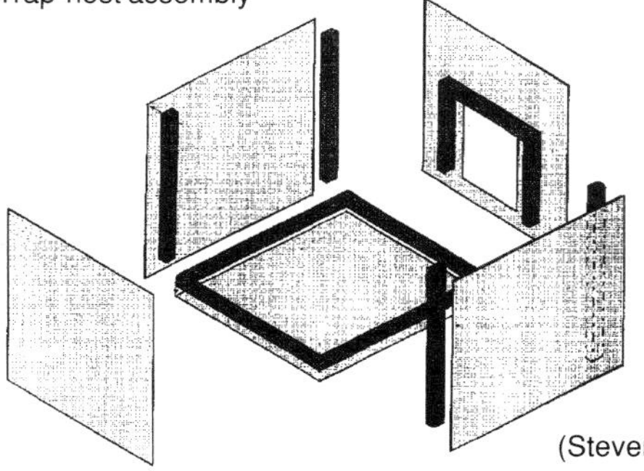

(Steven Rogers)

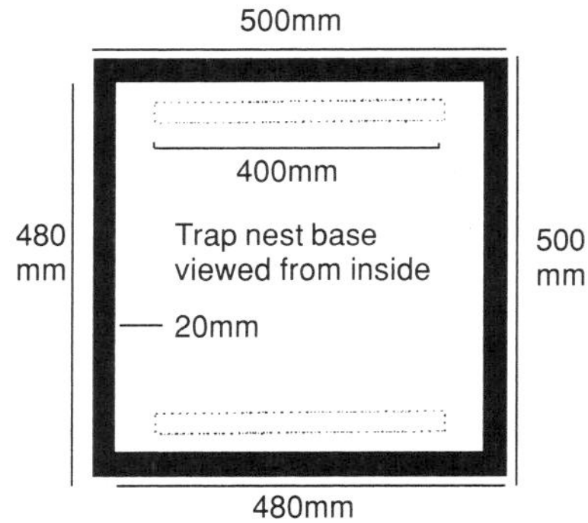

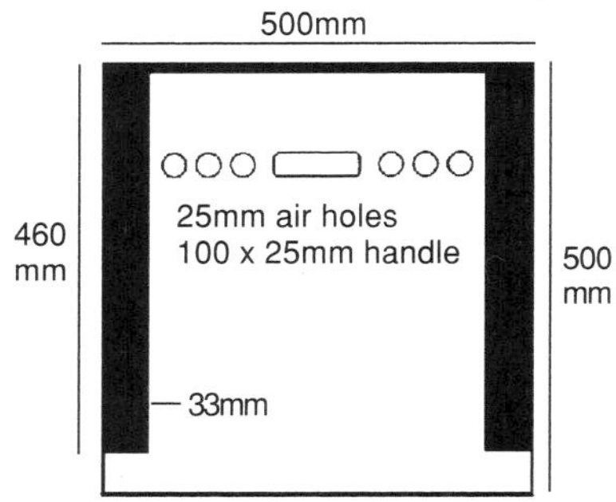

Trap nest viewed from outside

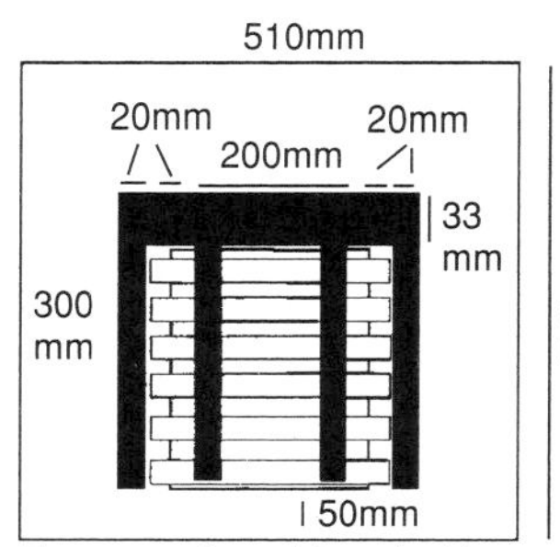

Trap nest viewed from inside

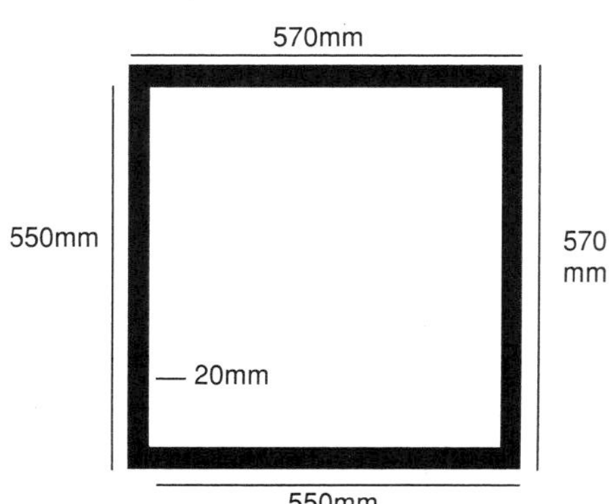

Trap nest lid viewed from inside

Breeding

There are several strategies when it comes to breeding:

Line breeding

This is breeding closely related birds so that a specific line or 'strain' can be established. Once established, though the line has been developed from common ancestors, it can be designed to avoid the future mating of very closely related birds such as mother/son, father/daughter, etc.

In-breeding

This is the mating of closely related birds such as brother/sister, father/daughter, mother/son. It may be carried out when there is a specifically valuable characteristic that is required. On the whole, however, it is best avoided because the inherent problem with in-breeding is the increased risk of genetic defects.

Out-crossing

This involves the introduction, from another breeder, of an unrelated bird of the same breed in order to improve a particular strain.

It is worth emphasising again, the value of keeping records when any kind of selective breeding is taking place, and also for every stage of incubation, hatching and brooding. Some suggestions for record cards are given on page 71.

Mating

How breeding birds are housed and managed will be largely a matter of individual preference. Some breeders have a permanent breeding house with alternate runs where the male and his females are kept permanently, using first one run and then the other to keep the ground from being overused. Others prefer to have relatively small breeding pens, such as the one illustrated on Page 19, which can be moved to new grass on a regular basis. The different systems are as follows:

Pen mating

This is where a small number of females are penned with one male. It is often the choice of the small breeder and the pedigree breeder, where the exact parentage of each chick must be known. There is usually one house for all the birds.

Flock mating

This is where a number of males run with a large flock of females. The extensive conditions ensure that the males do not fight. There are often several houses to cater for the whole flock.

Artificial insemination

Although practised within the intensive poultry sector, this is not a technique that is often found in the small sector. Nevertheless, it is possible to extract semen from the male by careful genital manipulation so that it is collected in a sterile container.

It can then be transferred to a sterile syringe and injected into the cloaca of the hen, at the base of the oviduct. Great care must be taken not to injure the hen. Best results are obtained when the fluid is transferred immediately.

It is a technique which is used with breeds of poor fertility - perhaps where there has been too much in-breeding in the past, or selection for appearance rather than for vigour. With this system, males and females can be kept separate, and the use of a male from another breeder at a different site is possible.

Feeding Breeding Stock

The best diet for breeding birds is a proprietary ration formulated for *breeders* of the particular species. Many poultry or waterfowl breeders give a ration of layers' pellets and grain. This is a perfectly adequate basic diet, and where they are able to forage on a reasonably extensive basis and obtain a proportion of live food, the birds may perform well. But, as a breeder ration it is not ideal for the production of healthy, fertile and viable eggs.

If you are finding that your hatches are consistently poor with a high proportion of weakly chicks, and there is no question of using closely-related or sickly parents, or of bad incubation practice, then you should suspect the parents' feed ration.

In a proprietary breeder ration there is an adequate balance of the necessary amino-acids, vitamins, minerals and trace elements. The problem for small flock owners is that breeder rations are not always easy to obtain in small quantities.

If you have only one or two breeding pairs of poultry, it is hardly worth the time, trouble and expense of buying in large quantities of expensive breeder ration which may well go off before you get around to using it all. Many small breeders use considerable ingenuity in making their own rations. One which I have used to good effect with chickens is as follows:

One crushed multi-vitamin tablet
One teaspoon medicinal cod-liver oil
One teaspoon wheat germ extract
One teaspoon Marmite
½ cup of milk
Enough layers' mash to mix to a coarse crumb consistency.

The multi-vitamin tablets are the sort which are manufactured for human consumption. These, together with cod-liver oil and wheatgerm extract, are available in most chemist and health food shops. This ration is enough for one breeding pair and is given once a week, in the morning, with grain in the afternoon. The rest of the week they are given normal layers' pellets and grain. The other important elements of their diet are access to fresh, clean water at all times and, of course, crushed oystershell and insoluble grit to ensure sufficient calcium levels, and to ensure that grain is broken down properly in the gizzard.

It is important not to overfeed (fat birds may not be too ready to mate) and not to incubate eggs from parents being treated with antibiotics. The effect of antibiotics is to kill the normal and useful bacteria of the gut, as well as those causing problems. When this happens, birds are not able to synthesize folic acid, with a resultant decline in the number of red blood cells. Where a breeder bird has had to be treated with an antibiotic for a specific infection, add a little plain yoghurt to its feed for about three days after finishing the antibiotic course, and avoid taking any eggs for incubation for another week after that.

Breeders of other birds such as ostriches, emus and rheas can obtain specialized rations from companies catering for a range of bird diets.

In the section *What went wrong?*, there is a list of the problems likely to be encountered with hatches from inadequately fed parent birds.

Influence of light on fertility and egg production

The availability of light has a crucial influence on fertility and egg production. In completely natural conditions sexual activity does not take place until spring returns with its lengthening days. The emu, laying between November and March, is an exception. Supplying artificial light to extend natural daylight will induce early egg production and the provision of fertile eggs. This enables breeders to produce much earlier hatched young than would normally be the case.

The most effective way of acquiring early fertile eggs is to provide artificial light after a period of short-day conditions. In other words, if egg production ceases in the autumn, as daylight dwindles, allow the breeding birds to experience this until the end of December. This not only gives their systems a rest, but enables you to get them into good condition for breeding, ensuring that all their nutritional needs are catered for. When extra light is provided, say from early January onwards, the message they receive is that 'it is time to start laying'. Young birds need a different lighting scheme. This is referred to in the section on rearing.

The time of year when artificial light is provided is, of course, a personal decision. It should be remembered that early hatched birds require much care and attention in relation to brooding temperatures because of outside winter conditions. Some people may prefer not to give artificial light at all and work within the seasons.

Providing artificial light is not difficult. A 25 watt bulb is sufficient for up to 25 birds. It can be powered by mains electricity or a 12 volt car battery. If a time switch is included in the circuit, the system can be programmed to switch on and off automatically, coinciding with the need to extend daylight from late afternoon onwards. A DIY system is not difficult to set up, for all the parts are available from motor and electrical suppliers. Alternatively, poultry equipment suppliers now supply complete lighting packages. The provision of 15 hours of light a day - a mixture of natural and artificial - will cover the needs of all birds in this respect.

Coturnix (Japanese) quail chick just emerging from its shell.
Poultry World.

The incubation area

Commercial hatcheries have three distinct areas - *storage, setter* and *hatcher* rooms - to cater for the three periods - *pre-incubation* (up to 7 days), *setting or incubating* until 2-3 days before hatching, and *hatching* when the final emergence from the shell takes place. Temperature and humidity of each area are kept at the optimum levels for the different stages. The following are the recommended conditions for the different hatchery rooms, not the incubator and hatcher themselves:

Room	*Air temperature*	*Relative humidity*
Storage	15-18°C	75%
Setting	24-28°C	60%
Hatching	24-28°C	60%

(Source: *Buckeye International Ltd*).

The small breeder is unlikely to have such distinct areas, and one room usually has to cater for all the activities. This may be a store room or even a spare bedroom, but every effort should be made to keep the environment stable, with no sudden fluctuations of temperature and humidity. In this way, there is less likelihood of fluctuations *inside* the incubator.

Many people do not realise the importance of this - that the outside environment can affect the inside of the incubator. A cold shed, for example, may have the incubator's heater and thermostat working flat out trying to keep the temperature stable, while cold, damp air can create havoc with the relative humidity.

A spare bedroom is much better than a shed. The insulation is usually adequate and central heating can be adjusted, via a thermostat, to keep the temperature stable, and to avoid draughts. If sunlight is too strong, it is always possible to draw the curtains. Ventilation is also sufficient, as a rule, with aeration only becoming a problem where large hatcheries are concerned. Here, fans are often used to duct the air through so that oxygen is replenished and carbon dioxide removed.

Having a stable background temperature helps to stop humidity becoming excessive. (It is much easier to increase humidity than it is to reduce it!).

It is also important to house the incubator where it is unlikely to be knocked and jarred. A corridor or frequently used room would not be appropriate, with all the household traffic passing through.

Finally, there is little point is washing and dipping eggs to ensure that they are free of pathogens, if the incubation area itself is not kept clean. Wipe surfaces regularly with an anti-bacterial preparation, and ensure that hands are treated in the same way before any operations such as candling are carried out. People often comment that the broody hen does not carry out such operations, but then she is not incubating as many eggs as even a small incubator, so there is less risk.

The incubator

An incubator, or *setter*, is essentially a container which provides optimum conditions for eggs while they are developing. They vary in size and capability, and in their power source. Most incubators, these days, work from mains electricity, although it is possible to obtain small machines which work from a 12 volt DC supply. Older machines used paraffin, and it is possible to buy these, although they need attention to detail such as keeping the burner wick trimmed to avoid smokiness, and so on. Nevertheless, the availability of both 12 volt battery and paraffin models is important in areas where mains electricity is not available. Gas is also used as a heat source, although it is not a popular choice with smaller breeders.

Choosing an incubator

In choosing an incubator, it is important to buy one which fits individual needs, and a mains electricity one would meet the needs of most small breeders. It is clean, convenient and, barring power-cuts, efficient. There are back-up electrical systems available however, so even the threat of a power cut need not be daunting. In areas of the world where mains electricity is not available, consider a paraffin (kerosene) model, or one that will run off a 12 volt rechargeable battery.

Posing the question, *'What is it needed for?'* is a good starting point because this will determine the size, type and ultimately the price of the incubator. If it is to be used for commercial purposes a reasonably large incubator will be required. Work out the number of eggs likely to be incubated, then add a margin for possible increase. Commercial breeders will generally opt for large, floor-standing setters and separate hatchers. For the smaller breeder, the main factors to consider when buying an incubator are as follows:

Table-top or cabinet?

If the number of eggs is fairly small, then the answer is probably a small *table-top model*. These generally have a hinged lid or lift-off access. Larger numbers would be more appropriately met by a *cabinet* incubator. This is one which has a door opening in the front, and has several shelves of incubation space.

If the incubator is for educational purposes, the ability to see into it without having to remove the top or open the door is an advantage. Some small incubators are designed in this way.

Separate setter and hatcher?

A *hatcher* is a container to which incubating eggs are moved just before hatching. This mitigates against the effects of carbon dioxide build-up and simplifies the hygiene routine. Other advantages are that the incubator area is kept clean, incubator space is maximised, and there is room for the chicks to move about during their

Parts of a table-top incubator

Powered: 220/230 volts x 150 watts - 110/120 volts

Dimensions: 275mm high x 353mm deep x 353mm wide
Weight: 12 kg

Curfew 236 - 20 chicken egg capacity

drying off period. Commercial hatcheries have separate setting and hatching areas. In this way, they are also able to provide the differing temperature and humidity levels required during the incubating to 'pipping' stage, and the hatching stage. Hatching needs a slightly higher humidity level and a slightly lower temperature.

For a commercial operation, or indeed any enterprise where maximum hatches are required, it is better to have a separate hatcher. If it is not an option, consider buying a cabinet model with a separate hatching drawer. Some small, table-top models have an insert that can be placed in the incubator, alongside the eggs, so that newly-hatched chicks can be put in there as they hatch.

Specialist use?

Most cabinet incubators are adaptable in that they can take drawers and inserts of varying sizes to cater for different birds.There are also some manufactured specifically for specialised use such as ostrich, aviary or zoo applications. Depending on the scale of operations, a specially adapted setter and hatcher may be appropriate.

Still air or fan-assisted?

The choice is now for one which is *still-air*, or *forced-air (fan-assisted)*. The former is where air circulates by convection, with warm air rising and displacing colder air, bringing about a natural circulation. In means that air flow must be worked out in order to ensure that there is not too much temperature fluctuation in different areas of the incubator. Still-air incubators are cheap, but less adaptable than forced-air models. However, on a very small scale, aeration will normally be sufficient if there is only one level of eggs, and the manufacturer's instructions in relation to opening and closing the ventilation holes are followed.

Forced-air or *fan-assisted* models are those where air is driven over the eggs by means of a fan. Large incubators need to have such a facility, otherwise there is insufficient aeration. Some small incubators are equipped with a built-in fan.

Manual or automatic turning?

Another choice to be made is whether to buy an incubator with an automatic egg turning facility, or one where the eggs must be turned manually. The very simplest of the table-top models require you to turn each egg individually and as this needs to be done two to three times a day, and ideally, five times, it is quite a commitment. If turning is neglected, this will be reflected in a reduced hatching rate. There are some incubator models available where individual eggs do not need to be turned; the whole batch is turned at the same time by an external control lever - but it is still manually operated. There are occasions, when a manually turned incubator may be preferable, as for example, in a school, where the children's participation is a necessary part of the educational project. (Remember to organise hatches so that they do not coincide with school holidays!)

Generally speaking, however, an automatic turning facility is preferable. Turning may be by a system of rollers, a 'moving carpet' or trays that are tilted first one way, then the other.

Temperature and humidity controls?

The heating element is normally provided by the electricity supply. A key part is played by the thermostat in controlling the temperature which must be kept at the optimum for the particular species. An electronic thermostat is much more efficient than the old wafer-type of thermostat, being more sensitive to temperature fluctuations and responding instantaneously, to switch the heat on or off.

The older type is much slower to react and a variation of 2-3 degrees either way is possible. This variation can be crucial to vulnerable embryos, and even on the smallest scale, an incubator with an electronic thermostat is a *must*.

An easy-to-read thermometer is also a necessity. These are either mercury, alcohol or digital. They can either be inserted from the side or top of the incubator, or are suspended just above the eggs. The crucial thing, as far as getting an accurate reading is concerned, is to ensure that the thermometer is placed close to the eggs, not high above them. Remember, however, that still air incubators may be colder in different areas. It may be necessary to have a higher overall temperature to ensure that the temperature at the eggs is correct. Follow the manufacturer's instructions!

Many incubators provide humidity from some sort of water reservoir which may need to be topped up manually. Some models use felts which need to be kept moistened, and it is essential to follow the manufacturer's instructions precisely. Most suppliers of general purpose incubators provide instructions which vary slightly depending on the type of eggs in question. An important point is that when water is added to the reservoir, it is warm water, not cold. The latter will cause a rapid drop in temperature with possibly damaging results. It is just as important, obviously, not to add water which is too hot. If in doubt, take the temperature of the water in the water tray and use that as a guideline, although the clean finger test in association with common sense is usually satisfactory.

Hygrometers can be used to measure the humidity but will not affect it. In order to reduce humidity, a separate unit or dehumidifier is now available. This can be used in conjunction with an existing incubator. Large setters often incorporate built-in dehumidifiers. More information is given on humidity in the section on incubation practice.

What construction?

The last factor to consider in choosing a small incubator is the type of material used in the construction. A well made model needs to be well insulated to minimize temperature fluctuations and heat loss. The material needs to be reasonably strong to stand up to long-term use, and should be easy to clean.

Plastic incubators are certainly easy to clean, but the insulation properties of low-grade plastics are fairly poor. High grade plastics are much better. Wood has reasonably good insulation properties, but can be difficult to keep clean. Plastic faced wood is a good compromise, and many incubators are made of this material. They are easy to clean, robust and well insulated.

Expanded structural polyurethane is strong and easy to clean. It also has good insulation value which has a direct effect on running costs.

Incubators which have a clear plastic top allow an unrestricted view, but may have reduced insulation qualities. This is not always the case, particularly where a double thickness is used. High quality plastics have good insulation properties.

Plain polystyrene incubators are inexpensive and have good insulation. However, they are not strong, unless they have been coated with tough plastic. Care should be taken in handling and cleaning uncoated polystyrene.

Finally, ensure that any electrical appliance you buy has BEAB safety approval or its equivalent in the country of use.

On the next pages are some of the incubators that are available:

Small Table-top incubators

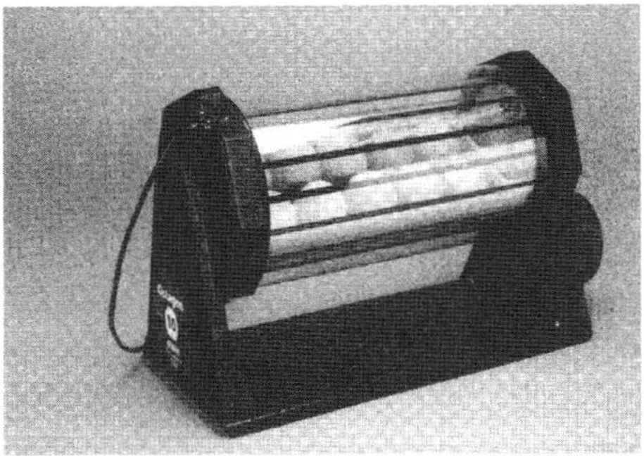

Octagon 10. Still-air, automatic turning, 10 eggs. *Brinsea.*

Octagon 20. Fan-assisted, 24 eggs, automatically turned. *Brinsea*

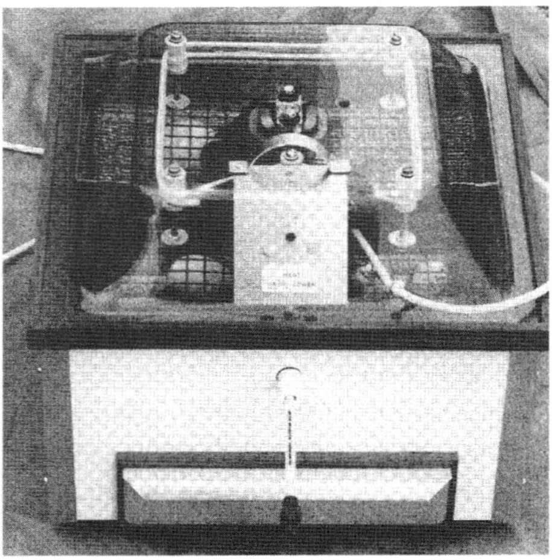

Curfew 236. Fan-assisted, automatic turning, hatching tray, 20 eggs.

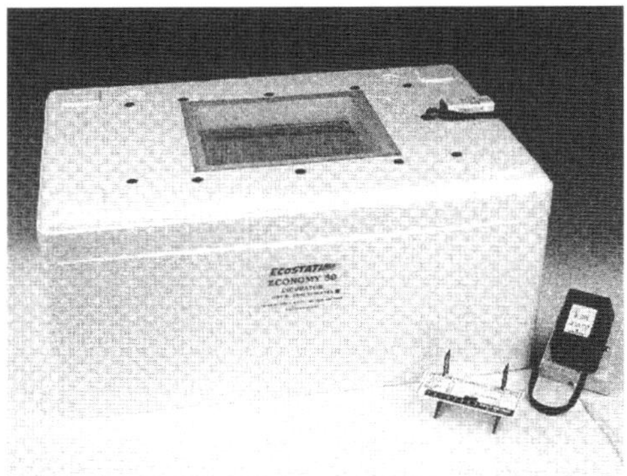

Economy 80. Still air, manually turned, 80 eggs, available in 12 volt and 230 volt. *Ecostat.*

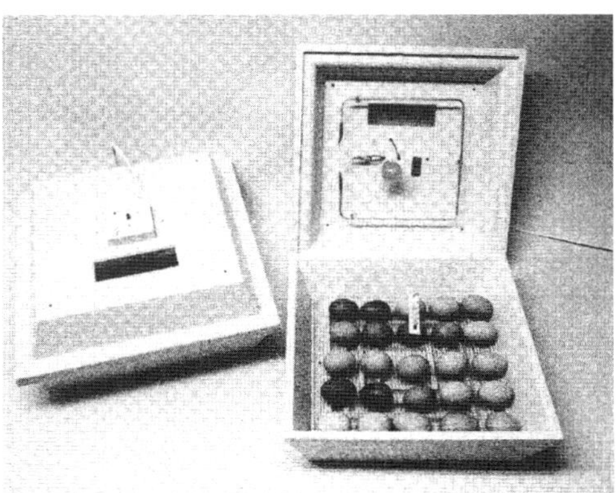

Therbo. Still-air, 25 eggs, available as semi-automatic or automatic. *Torne Valley*

Covattuto 12. Still-air, manually turned, 12 eggs. *Southern Aviaries*

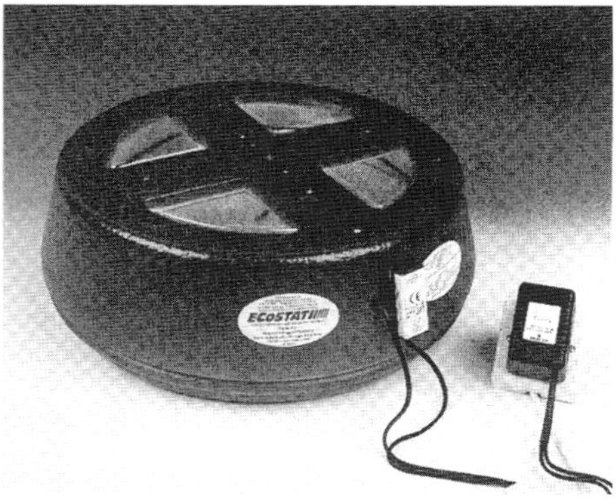

Ecostat. Still-air, 60 eggs manually turned or 30 eggs with semi-automatic mechanism. *Ecostat*

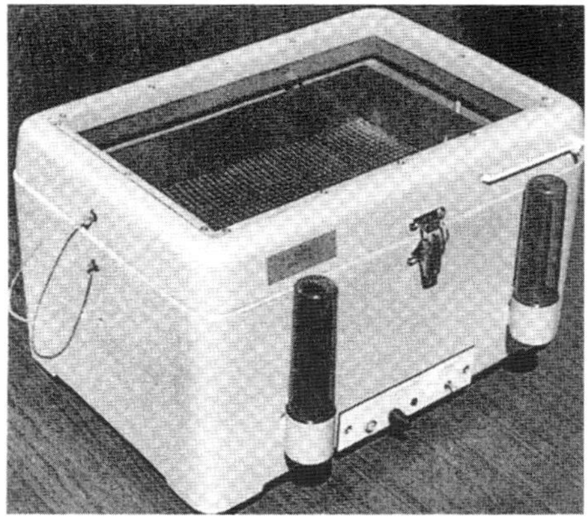

A.B. Startlife 25 hatcher. One of a range for waterfowl and exotics. *A.B. Incubators*

Lyon TX6. Fan-assisted, automatic incubator with humidity controller. *Lyon*

Polyhatch. Still-air, automatic, 42 eggs. Available for use with mains electricity or 12v DC version. *Brinsea.*

Matador Professional. Fan-assisted, automatic, 50 eggs. Suitable for parrots and exotics. *Torne Valley*

FIEM MG25. Still-air, semi-automatic, 25 eggs. *Norfolk Game Supplies*

(Unless stated differently, egg size refers to chicken eggs)

Cabinet incubators and hatchers

Fiem MG 50. Fan-assisted, semi-automatic or fully automatic, 55 eggs, *Norfolk Game Supplies.*

MS Broedmachine. Fan-assisted, automatic and semi-automatic, 35-50 eggs. *MS Incubators.*

NatureForm NOM 50/90. Fan-assisted, automatic, designed for ratites. 50-90 ostrich, 120 emu, humidity controller. *NatureForm.*

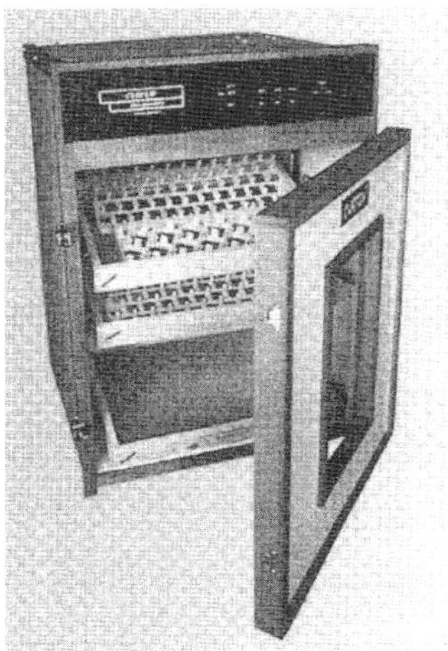

Curfew 247. Fan-assisted, automatic, 100 eggs.

Masalles 1200-25. Fan-assisted, automatic, 288 eggs. *Banbury Cross Veterinary Farm Supplies*

Miracle. Fan-assisted, automatic, separate setting and hatching trays, 3 models - up to 224 eggs. *Torne Valley.*

Combination setter/hatcher with incubation trays above and hatching baskets below. Fan-assisted, automatic, humidity controller, 2376 eggs and 792 chicks. *Bristol Incubators.*

Marcon RS8000 large incubator. Fan-assisted, automatic, humidity controller, suitable for large scale incubation. Here, it is fitted with trays for ostrich eggs.

From start to finish

Stages in development of the chick embryo

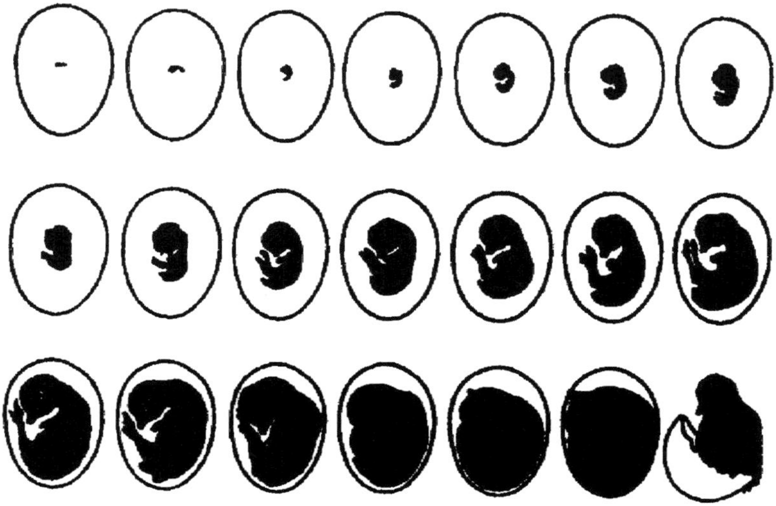

Day	*Development*
1	Beginnings of digestive canal, spinal column, nervous system, head, eyes.
2	Beginnings of heart and ears.
3	Heart begins to beat. Formation of legs and wings.
4	Formation of tongue.
5	Reproductive organs begin to form.
6	Formation of beak.
8	Feathers begin to grow.
10	Beak hardens.
13	Formation of claws.
14	Takes up position for eventual emergence.
16	Final hardening of beak, claws and leg scales.
17	Beak placed into position by air sac.
18	Yolk sac begins to be absorbed into the abdomen.
19	Pipping commences.
21	Hatching.

Incubation principles

In order to incubate successfully, a fertile egg needs the following:

- to be clean, undamaged and free of pathogens.
- stored at a temperature of 15-18^{0}C and a relative humidity of 75%. (If it is to be stored for more than 7 days before incubation, reduce the temperature to 12^{0-}15^{0}C).
- ideally to be no older than seven days.
- incubator temperature of 37.5^{0}C at the centre of the egg, although there are slight variations depending on type of birds. (See the Table on page 49).
- incubator relative humidity of 30-60%, depending on the type of eggs. (See the Table on page 49).
- hatcher relative humidity of 75%.
- adequate aeration from good ventilation.
- regular turning - ideally five times a day, unless it is automatically turned.

Where these conditions are available, the egg will develop according to the pattern indicated opposite. This is the pattern shown by the chicken embryo. Although the relative time scale may differ for different types of bird, the pattern of development is essentially the same.

Hygiene

Reference has already been made to the importance of good hygienic practice. If you wish to avoid disease problems, wash dirty eggs in warm water and a suitable egg sanitant from incubation suppliers. Dip them again before incubation if there is a delay. Thoroughly clean and disinfect the incubator and all surfaces where the eggs may be placed. Wash hands with an anti-bacterial product and continue to observe hygienic practices at every stage.

See the section *What went wrong?* for details of problems that can arise from inadequate or non-existent hygienic practices.

Temperature

The ideal temperature requirement for incubation of most eggs is 37.5^{0}C at the centre of the egg. With some still-air incubators, the temperature may vary in different parts of it. Check this carefully! The reading may need to be 39.5^{0}C two inches above the eggs to ensure that it is 37.5^{0}C at the centre of the eggs. Be guided by the specific instructions that come with the incubator.

In a forced-draught, fan-assisted model, the temperature is generally the same in all areas of the containment area. The thermometer should therefore read 37.5^{0}C.

It is a good idea to buy several thermometers, not only to mitigate against the effects of breaking one, but also to run checks at different levels. A maximum-minimum thermometer is also useful for running a check on the incubator while it is being set up. It should, of course, be allowed to run for at least 24 hours, and ideally 48 hours before use.

As the embryos develop, they generate their own heat so that this, added to the existing heat, increases the overall temperature. In small incubators, where ventilation is adequate, this is not normally a problem for the surplus heat disperses quite easily, by opening the appropriate vents. Even so, keep an eye on the thermometer in case of temperature fluctuations and, if necessary, increase the ventilation.

In large hatcheries, special cooling coils have to be incorporated to prevent the temperature becoming too high. Temperature fluctuations can kill or damage the developing embryos, with the most critical period being the first few days of incubation. Towards the end of the incubation period, fluctuations can be more easily tolerated, as long as they are not too extreme. As a general rule, a slight drop in temperature is less critical than an increase at this stage.

Ventilation

Adequate ventilation is essential. Oxygen from the incoming air is needed by the incubating embryo. This diffuses through the shell pores into the air sac and into the tissues and blood of the embryo, with carbon dioxide and water vapour being removed in the same way. As the embryos develop, their oxygen requirements also increase, so that provision for an increased air flow is necessary in incubators. With a small, still-air incubator, this is simply a matter of opening more of the ventilation holes, and the manufacturer's instructions are crucial in this respect.

People living at high altitudes where the air is thinner, with a reduced oxygen content, often have problems in incubating eggs satisfactorily. The solution here, is to increase both ventilation and temperature slightly. Fan-assisted incubators have a power-assisted system of ensuring an even flow of air over the eggs.

When a DIY incubator is constructed, provision for extra ventilation holes is important, and the problem, of course, is achieving the correct balance between acceptable air-flow and an adequate level of insulation. As a rough guide, a small table-top model would need to have a complete air change every hour. An easy way of checking this is to puff some smoke into the incubator (from a bee smoker or a cigarette if you are a smoker). From being smoke-filled, it should be completely clear within an hour. Achieving this will be a matter of adjusting the ventilation holes accordingly.

Humidity

Moisture in the air ensures that there is enough water in the egg to keep the internal membranes from drying up, and also to ensure efficient diffusion of oxygen and carbon dioxide. If there is too much, the embryo does not receive enough oxygen and dies, a condition generally referred to as 'dead-in-shell'. Relative humidity generally needs to be around 50% for the incubator and 75% for the hatcher. (There are, of course, slight variations depending on type of eggs).

It used to be easy to maintain this level of humidity - usually just by keeping the water reservoir in the incubator topped up - while the temperature and aeration of the incubator stopped it from going too high. In recent years, however, climatic changes have increased the overall level of humidity in the air in many regions. Thanks to the 'greenhouse effect' it is now more difficult to keep the humidity level stable in the incubator, unless you can control it in the room in which the incubator is housed. The temperature does not keep it in sufficient check. The outhouse where I used to have my incubator frequently shows readings of 90% humidity. I now have my incubator in a small room inside the house where it is warmer and drier. It is more stable than before and hatches have improved considerably. A friend told me that a breeder he knows puts copies of *The Financial Times* in the Aga and when the pages are really warm and dry, puts them in the incubator to absorb the excess moisture and reduce the humidity!

The level of humidity is reflected in the relative size of the air cell in the egg. As the embryo grows, the air cell needs to enlarge in order to provide sufficient levels of oxygen for it breathe. Candling the egg against a bright light indicates the size of the air cell. Comparing this with the size it ought to be at a particular stage of development shows whether the relative humidity is correct. If the air cell is too small, the humidity is excessive; if too large, it is insufficient.

An even more accurate way of keeping a check on the relative humidity is to weigh the eggs at regular intervals. If all the eggs are approximately the same size and weight at the start, it may only be necessary to weigh one or two from the batch. However, with large, expensive eggs the practice is to weigh every one.

The eggs of poultry and waterfowl need to lose 11-13% of their initial weight up to the point at which egg turning ceases and they prepare to hatch. (For ratites it is nearer 13-15%). There is a formula for working out the percentage of weight loss, and this is indicated on page 43.

Dehumidifiers which will extract moisture from the air are available, and these can be used in association with incubators. Some are designed to sit underneath an incubator. Large incubators often have them built in, but even small incubators can be linked up to a dehumidifying unit so that the humidity can be reduced if needs be.

It is also possible to have scales which are linked to a computer. As the weight is recorded the data is automatically updated, with the appropriate compensation taking place in the dehumidifying unit, if the relative humidity is not what it should be for the particular stage of development.

This degree of technology is probably not appropriate for most small breeders, but as the use of computers becomes more widespread, and the cost of software falls, it may not be too long before the average family computer is used in this way.

Egg candling

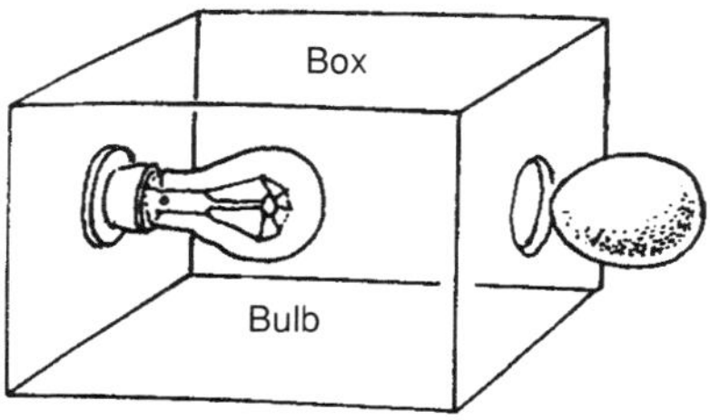

A home-made candler

Egg candled at 7 days

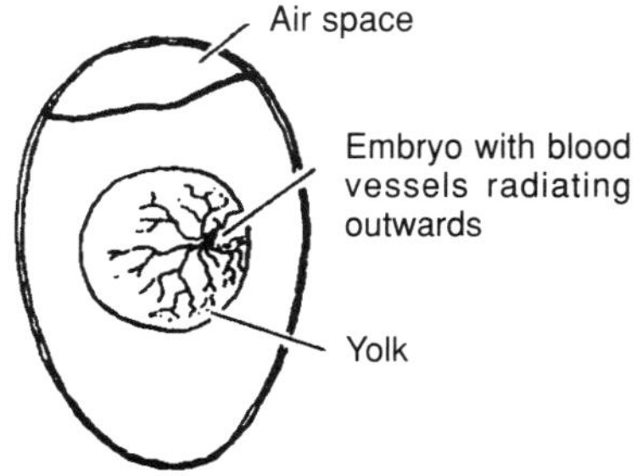

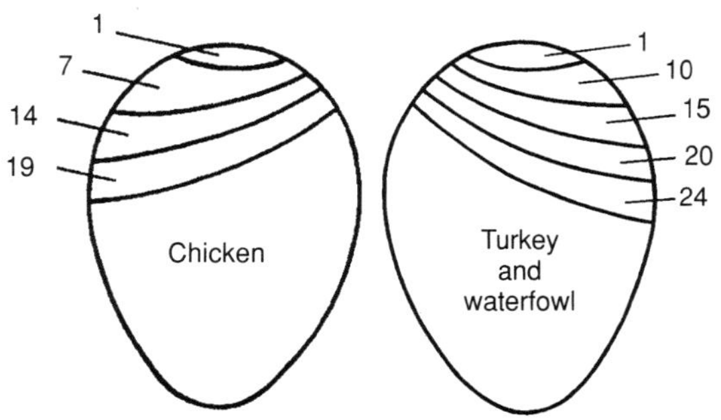

These show the size of the air cell that eggs should be at the appropriate number of days into incubation. It is an accurate way of determining whether the humidity level is correct because the amount of water is linked to the size of the cell. If it is too small, the humidity is too high. If too big, the humidity is too low.

Eggs can also be weighed and the weights recorded on a graph. In this way, if an expected weight loss of 11-13% is expected up to the pipping stage, the *expected* weight loss and the *actual* can be plotted on the graph, showing up any deviations from the norm.

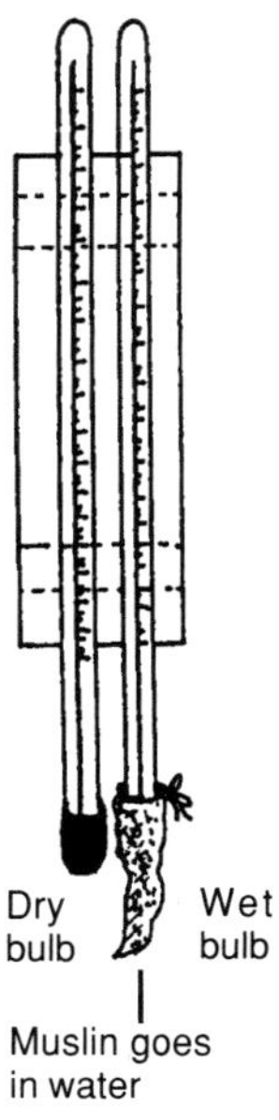

A home-made hygrometer

Tape two thermometers to a piece of plywood and tie a piece of clean muslin round the bulb of one. This is the 'wet' bulb while the other is the 'dry' one.

Place in the incubator so that the muslin is in the water but the bulbs are clear. The temperature of the 'dry' thermometer will be that of the incubator. The 'wet' bulb one will have a lower temperature because as water evaporates from the muslin it uses the latent heat in its surroundings.

Subtract the wet bulb reading from the dry bulb reading. A large difference shows that the atmosphere is dry, while a small difference indicates that humidity is high. The table on the right enables the readings to be translated into the relative humidity percentage, ie, the amount of water in the air.

Humidity measuring table using wet and dry bulb thermometers

Dry bulb temp - C	Wet bulb temperature - Centigrade		
	50% RH	60% RH	70% RH
37	28.3	30.3	32.3
37.2	28.8	30.6	32.4
37.5	29.0	30.9	32.8
38.1	29.2	31.2	33.2
38.7	29.7	31.8	33.9
39.3	30.2	32.3	34.3

To check the relative humidity read the temperature on the dry bulb thermometer and find its nearest reading in the first column. Now check the temperature of the wet bulb thermometer and find its nearest reading in any of the remaining three columns. If the dry bulb temperature is 37.5 and the wet bulb one is 32.8, the relative humidity is 70%.

If this seems too much like hard work, most incubator suppliers sell hygrometers to measure relative humidity.

The *550 humidity controller* - a dehumidifying unit for reducing excess moisture in the incubator. This model is suitable for 15-30 eggs. The incubator is placed on top. *Curfew.*

Rollers are used to turn the eggs in this table-top incubator. It is important to place the thermometer very close to the eggs to get an accurate reading.

Calculating percentage weight loss

The following formula can be used to work out the percentage weight loss of an egg (or batch of eggs) during the incubation period up to the pipping stage.

$$\%\text{ weight loss at (day 10)} = \frac{(\text{Day 0 egg weight} - \text{Day 10 egg weight})}{\text{Day 0 egg weight}} \times 100$$

$$\text{Average daily loss} = \frac{\%\text{ weight loss (Day 10)}}{10}$$

Projected 19 day loss = average daily loss x 19

So, if an egg weighed 60g before starting incubation and 58g on day 7, the % weight loss at day 7 would be $\frac{60-58}{60} \times 100 = 3.3\%$

The average daily loss would be $\frac{3.3}{7} = 0.47\%$

Projected 19 day loss would therefore be 0.47 x 19 = 8.93%

In this case, 8.93% is not enough of a projected weight loss, as it should be in the 11-13% range for adequate air cell development to have taken place. Action should be taken to reduce the humidity in the incubator.

A dehumidifier such as the one shown above can be used or the amount of water in the water tray can be reduced. If two water containers are used, remove one of them.

Incubation practice

Everyone has their own way of doing things, but the following is a recommended routine for the practical tasks associated with incubation and hatching.

Preparing and testing the incubator

Place the incubator in a protected environment, as discussed earlier, and check that all the necessary equipment is to hand. These include: scales, candler, antibacterial spray, spare thermometers, hygrometer (for measuring humidity), record book or chart, pens and pencils. You may also wish to invest in good quality rubber, surgical gloves and a nose mask. Remember to hang a thermometer and hygrometer in the room itself so as to keep a check on the overall environment, which is just as important as conditions inside the incubator. Check that all the thermometers give the same reading before starting.

Make sure that there are enough electrical sockets to plug everything in. If an incubator has a fan, it often has a separate cable and plug for this. Then there's the candler and there may be a dehumidifier, and so on.

Thoroughly clean and disinfect the incubator, then assemble all the parts. Make sure that roller bars or other turning mechanisms are correctly placed for ease of turning. Follow the manufacturer's instructions precisely. Take care not to splash too much water around electronic controls when cleaning!

Fill the water reservoir (follow the manufacturer's advice), then set the incubator going for at least 24 hours, preferably 48 hours. This gives an opportunity to check that the thermostat is operating properly, as well as ensuring that the optimum conditions are achieved before introducing the eggs. Check the temperature at different levels and only when you are satisfied that all is well, introduce the eggs.

Introducing the eggs

While the incubator is running through its preparatory run, the eggs should be brought from their storage conditions and kept at room temperature overnight. This will ensure that the sudden temperature change, when incubation starts, is not too drastic. If turning is to be manual, mark each egg with an **X** so that turned and unturned-turned eggs do not become mixed up. If they are from hens with numbered leg rings, it is appropriate to mark each egg with the mother's number.

In small incubators, eggs are normally placed on their sides. In larger ones, they are usually placed upright, blunt-end up in the supporting trays. If they were put sharp-end up, the chick's head would be away from the air cell (which invariably forms at the blunt end) and it would die. Where eggs are placed upright, the trays are automatically tipped at an angle of 45^0, first one way, then the other.

When the eggs are first introduced, the temperature may drop slightly in the incubator. Resist the temptation of turning up the thermostat! Once the eggs have warmed, the incubator temperature will stabilise.

It is not recommended that eggs of different birds are incubated at the same time, for their requirements differ. There is no reason, however, why the eggs of different breeds of the same species should not be included. Although the eggs can be marked differently - one lot with a black cross, one with a red cross, for example - once the eggs have hatched, it will be impossible to distinguish them. The solution is to separate the eggs into their own containers, such as a wire mesh insert, setting tray or box - or even a netting bag, although I would not recommend these.

The chicks are thus kept separate until they can be marked and placed in a brooder. An easy way of identifying them is to put a soft felt-tip mark on the top of the head, or somewhere on the plumage. Alternatively they can be wing-tagged or toe punched. Later on, plastic leg rings can be used when their legs are a bit bigger.

The point has been made earlier that eggs of the same size and shape will also tend to produce more uniform hatches. Over-large eggs may be difficult to fit into the trays.

Regular egg turning

Turning the egg does several things:

- It allows the nutrients to be brought into contact with the germinal disc, at a stage when the embryo does not yet have a blood system to transport nutrients.
- The developing embryo will become stuck to the shell membrane if the position of the egg is not changed regularly.
- The ability of the internal membranes to extract waste products produced by the chick becomes restricted with inadequate turning, causing a build-up of toxic by-products which eventually poison the embryo.

A broody hen turns her eggs frequently before settling down again on her nest. Small table-top incubators will require hand-turning of the eggs, at least three times a day, and ideally, five. The odd number ensures that the eggs are not always on the same side during the long night period. Always remember to wash your hands before handling the eggs in order to reduce the possibilities of introducing infection.

Some small incubators have a handle outside which enables you to turn the complete batch in one go. It is still manual turning, but an improvement on individual egg turning. The ideal is the incubator with an automatic turning facility. This is standard in large commercial incubators and is now available in small models as well.

Candling

Candling is a method of examining an egg against a bright light in order to establish what is happening inside the egg. The term is a reminder that candlepower was originally used. These days, purpose-made mains electricity or battery-operated models are available. There are table models as well as hand-held ones, or you can make your own candler. Some suggestions are shown on page 42.

Candling can be used to check that there are no hairline cracks or double yolks in the process of selection and assessment for incubation. At around 5 - 7 days into incubation, candling will establish whether or not the embryo is developing properly. If not, the egg is infertile and can be taken out to avoid 'going off' and possibly infecting the rest. At around a week, the embryo can be seen as a blob with blood vessels radiating outwards, rather like a starfish.

The size of the air cell also indicates the relative humidity situation, as discussed earlier. If it is too small, the amount of moisture present is too great, and humidity of the incubator needs to be decreased. If it is too large, excessive loss of moisture is occurring and the relative humidity of the incubator should be increased. These measures will include providing more or less water in the trays, as required, or using a dehumidifier. Where several incubators are run, it may be appropriate to move some eggs to another incubator if that happens to have a different and correct level of humidity for the eggs concerned. As discussed earlier, carrying out regular weighings is an accurate way of checking air cell development.

Light shelled and unpigmented eggs are relatively easy to candle. I have always found Maran eggs and quail eggs difficult to candle because the shell markings obscure the egg contents. On a small scale, candling may not be as crucial as it is with larger units, but there is still a risk of bacterial infection spreading to the rest of the eggs if there is one infertile egg which gradually goes off in the warm conditions.

Hatching

About three days before the chicks emerge from their shells, egg turning should cease because they will soon start to 'pip'. Pipping is the initial breaking through of the chick's beak into the air cell where it will have its first lungful of air, and then the piercing of the outer shell. The beak has a specially adapted area called an 'egg tooth' which makes this easier, and for about a day before, the chick will have manoeuvred itself into the correct position. So, it is important to stop turning the eggs otherwise the poor chick will be completely disorientated.

Commercial breeders transfer the trays or trolleys of eggs from the setters or incubators to separate hatchers. Here they have a higher humidity and a slightly lower temperature. The transfer also makes the control of carbon dioxide easier, as well as simplifying hygiene.

Those with small cabinet incubators may have a separate tray to which the eggs can be transferred before pipping. Some table-top models have a separate tray insert. Whatever equipment is used, the salient points are to reduce the temperature slightly so that humidity increases. Increase the amount of water if necessary.

For the last couple of days before hatching, the chicks can be heard cheeping

Mark eggs with a cross if they are to be manually turned.

inside the shells. With a broody hen, they are given positive encouragement to hatch because she cheeps to them in return.

Although there is an average incubation period for each type of bird, there are always some which incubate early, and others late. This is normal, and it is important not to switch off the incubator too early, in the mistaken belief that everything which was going to hatch has done so. Allow at least three days over the average period, to take late arrivals into account. As far as helping individual chicks out of the shell once they have pipped, there is a considerable conflict of ideas. Some people state that no help should be given; others maintain that it should. My own view is that if an individual chick has pipped and then little or no progress is made in the next few hours, help should be given, but carefully! Pick up the egg and examine the membrane around the hole. If it looks dry, then get a flannel dipped in warm water to which a little egg sanitant has been added, and wrap the egg in it for a moment. Then gently enlarge the hole by picking a little of the shell on either side, and cut a little more of the membrane. If it begins to bleed, stop immediately, and replace the egg in the incubator. It is not ready!

If there is still no progress a few hours later, repeat the process, by which time, enough of the shell and membrane will have been weakened to enable the chick to emerge alone.

The author's son, Gwilym, inspects some newly hatched Aylesbury ducklings.

Drying off

Leave the hatcher undisturbed while the chicks dry off. The down feathers are quite wet when they emerge, but soon dry off and fluff up. During this period, they are obviously susceptible to temperature fluctuations, and the door of the hatcher should be kept closed.

They can spend 24 hours before being transferred to a brooder, giving them a chance to find their feet and walk around. The remnants of the yolk are enclosed in the abdomen, ensuring that they have enough food to keep them going during this period. Any physically damaged or congenitally deformed chicks should be quickly and humanely killed. It is far more inhumane to keep deformed chicks alive, in the vain hope of restoring them.

Remove all the empty egg shells and place in a thick plastic bag for disposal. Dismantle the hatcher, then thoroughly clean and disinfect it so that it is ready for the next batch of eggs. Make sure that all the details of the hatch have been recorded. On page 71 there is a suggested incubator record card, while page 94 gives details of how to work out the fertility and hatching percentages. Further details on how to care for the chicks are given in the rearing section.

Optimum conditions for incubation and hatching

Bird	Temperature °C		Humidity RH%		Pipping	Hatching
	Incubator	*Hatcher*	*Incubator*	*Hatcher*		
Chickens	37.5	37.0	52	75	18 days	21 days
Ducks	37.5	37.0	58	75	25	28
Muscovy	37.5	37.0	60	75	31	34
Ornamentals	37.5	37.0	55	75	19-27*	22-30*
Geese	37.5	37.0	55	75	28	31
Chinese	37.5	37.0	55	75	27	30
Light	37.5	37.0	45	75	27-30*	30-33*
Heavy	37.5	37.0	50	75	31-33*	34-36*
Turkeys	37.5	37.0	55	75	25	28
Guinea fowl	37.5	37.0	55	75	25	28
Quail						
Coturnix	37.5	37.0	45	75	15	18
Bobwhite	37.5	37.0	45	75	20	23
Chinese painted	37.5	37.0	45	75	12	16
Pheasants						
Game	37.5	37.0	50	75	21	24
Ornamental	37.5	37.0	50	75	20-25*	23-28*
Partridges	37.5	37.0	47	75	20	23
Peafowl	37.5	37.0	50	75	25	28
Pigeons/Doves	37.5	37.0	50	75	12-13*	15-16*
Swans	37.5	37.0	50	75	26-32*	30-36*
Ostrich	36.0	35.5	30	75	35	42
Emu	36.0	35.5	40	75	46	50-52
Rhea	36.0	35.5	40	75	33	36
Parrots	37.5	36.6	50	75	15-26*	18-29*

* Varies depending on breed.

A small incubator made with an Ecostat incubator kit. The buyer provides his own box which can be built following instructions that come with the kit. *Ecostat*

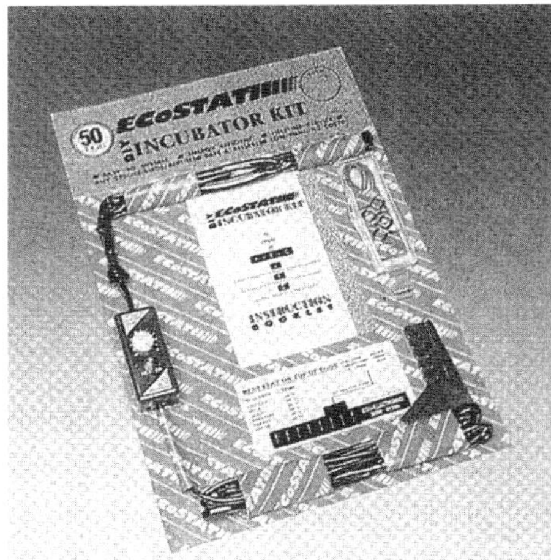

The Ecostat incubator kit, available in four sizes, depending on the egg capacity of the box to be used. *Ecostat.*

The Eco-element is mounted on 40mm lengths of dowelling fitted to the top of the incubator cabinet.

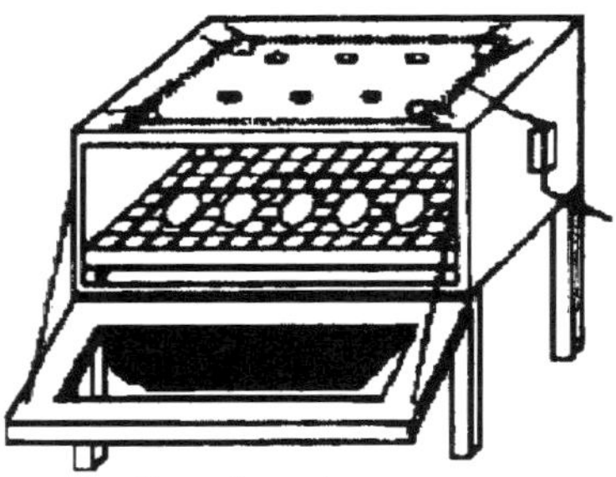

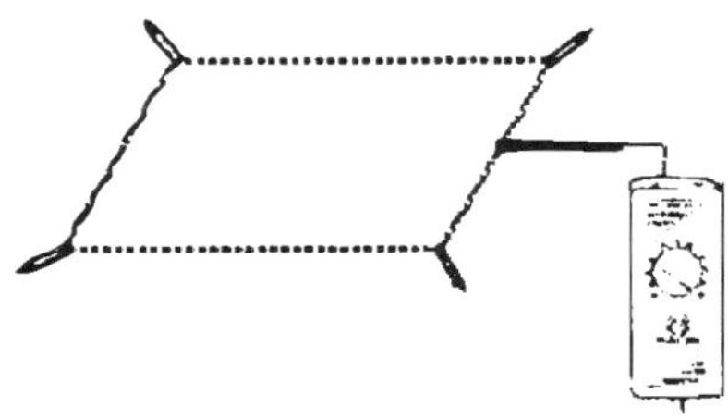

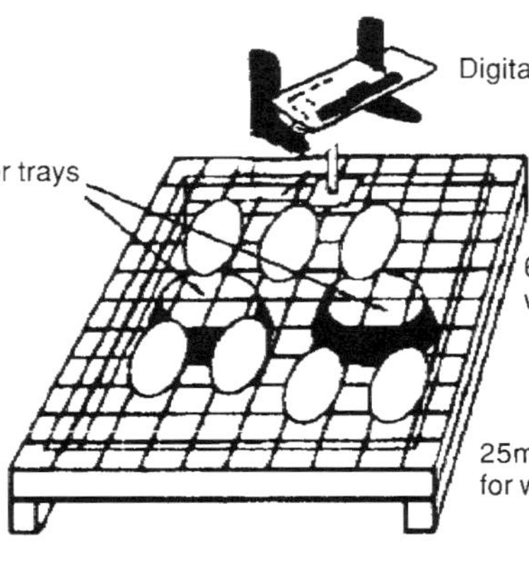

(By courtesy of *Ecostat*)

Home-made incubators

Making your own incubator is feasible, although care is needed with the fine adjustments. It is a question of being able to provide conditions where temperature and humidity requirements are balanced against the need for ventilation. The following illustrate some ideas which have been successfully put into practice.

The Ecostat incubator kit

One of the best ways of making your own small incubator is to use a kit containing the necessary electronic components. DIY kits are available from *Ecostat* in four sizes. These are suitable for 25, 50, 75 or 100 chicken egg capacity. Each kit consists of an instruction leaflet, electronic thermostat and thermometer, together with heating element and sensor. It is assumed that the purchaser makes his own box, although recommendations on the best construction are included.

The following dimensions of box are suitable for the various capacities:

25 egg	Length 350mm x Width 230mm x Depth 180mm	25 watt heater
50 egg	350mm x 460mm x 180mm	50 watt
75 egg	525mm x 460mm x 280mm	75 watt
100 egg	700mm x 460mm x 280mm	100 watt

9mm plywood is recommended by the supplier, with 25mm polystyrene for insulation. The latter can be bonded to the plywood with silicone sealant, and may need periodic replacement. Alternatively, another layer of plywood can be added to 'sandwich' the polystyrene and simplify cleaning. One side has a downward-opening door with glass observation window. Double-glazed glass or perspex will ensure adequate insulation.

For ventilation there are 6mm holes in the roof and floor. Four holes are needed for the 25 and 50 egg capacity models, and 6 for the 74 and 100 egg sizes. The incubator should stand on small feet to keep it clear of the ground, in case ventilation is impeded. Inside the incubator, 6mm twin weldmesh is used for the egg tray, with 25mm supports to keep it clear of the bottom. It should fit snugly in the box, with no gaps at the side. Plant pot coasters can be used under the weldmesh to act as water containers. One is sufficient for incubation, two for hatching. (A 50mm sized coaster is suitable for the the 25 egg size; 62mm for 50 egg; 75mm for 75 egg and 87mm for 100 egg).

Four 40mm lengths of dowelling rod are used to mount the heating element attached to the roof of the box. The element and sensor are mounted through a 14mm hole on the side of the container with a clip provided.

A 13amp 3-pin plug fitted with 2 amp fuse is attached and the incubator run at a temperature setting of 39°C (which ensures 37.5°C at the centre of the egg). The thermometer is placed in the tray where it can be seen through the window.

The Home Farm incubator

Dimensions are given in inches rather than in metric measurements because that is how I made it.

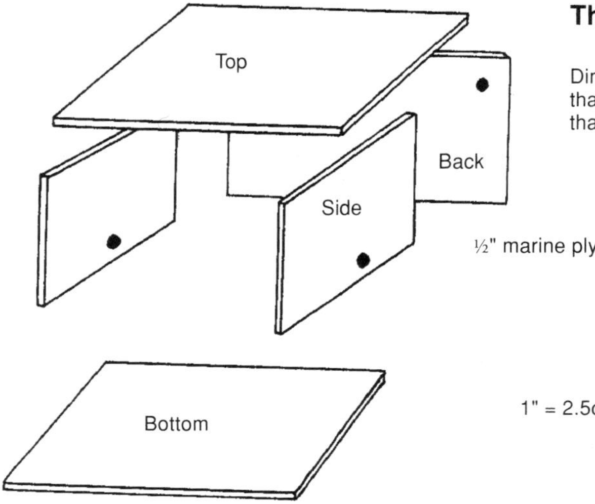

1" = 2.5cm or 25mm

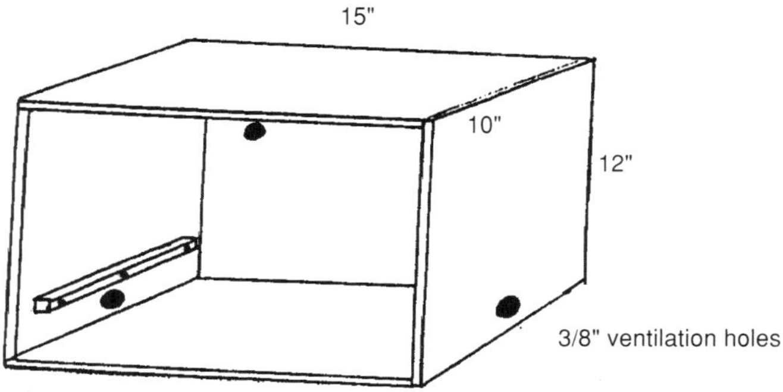

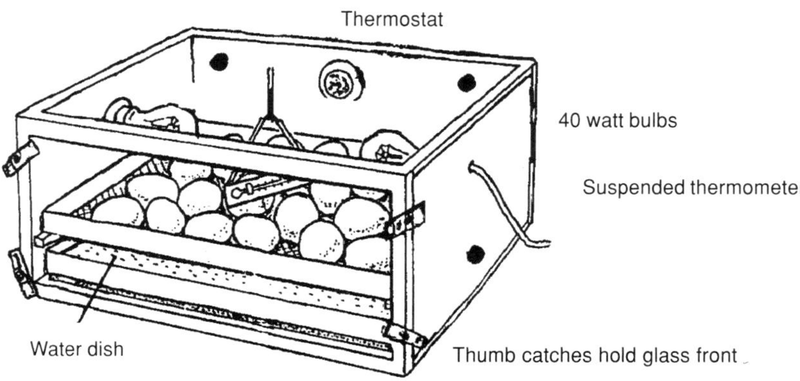

The 'Home Farm' incubator

Some years ago I built a small incubator, with help from my husband, David, for my knowledge of carpentry is limited. After a few false starts, we produced a model which subsequently hatched chicken, duck and quail eggs. Details of the incubator were published in *Home Farm* magazine (now *Country Garden & Smallholding*), hence the name.

Materials

Imperial dimensions are given because when we first built this it was in pre-metric days. Convert to the nearest appropriate measurement, bearing in mind that 1 inch is 26mm.
2 pieces ½" marine ply 10" x 15" (for top and bottom)
1 piece 12" x 15" for back
2 pieces 10" x 12" for sides
Sheet of glass 12" x 15" for front
2 wooden runners 9" x ½" x ½"
2 strips of wood 14¼" x 1" x ½"
2 strips of wood 9" x 1" x ½"
4 thumb catches 1 aluminium baking dish
Piece of twilweld galvanised wire mesh 14¼" x 9" x ½" gauge
Two 40 watt bulbs
1 suspended thermometer
1 modeller's thermostat
2 pieces thin foam rubber insulation 15" x 10"
Insulation tape
Sockets for bulbs, with cable and plugs.

Four ventilation holes, $^3/_8$" in diameter, were drilled in the marine ply, two at the top of the back section, and one on each side piece near the bottom. As the sheet of glass which I had as the front section was not particularly well fitting, I felt that more ventilation holes were unnecessary. If yours is close-fitting, it may be a good idea to have another two ventilation holes. Trial and error is valid here; if you find that you have made too many holes, tape one of them over.

The foam rubber insulation was stuck onto the bottom section, and onto the top one. (The bottom one did eventually prove a nuisance to clean, and I subsequently removed it, replacing it with a new piece of aluminium cooking foil each time the incubator was used.) The top, bottom, side and back sections were screwed together, and insulation tape used to provide added seals along the joints. A bulb socket was screwed to each side, with a hole made for the cable to exit. The thermostat was screwed to the back forming part of the circuit with the light cables. The two runners were screwed into the sides of the incubator, allowing enough room below for the water dish to be pulled in and out. A little vaseline was rubbed onto the top surface of the runners so that the egg tray could slide in and out easily. The egg tray itself was made by screwing together the wooden strips and then tacking on the wire gauze. A piece of strong wire was tacked to the roof and formed into a hook. The length of this was adjusted so that when the Elt thermometer was suspended, it was exactly 2" above the middle of the egg: Finally, four thumb catches were attached to the front so that the glass was held in place. Once the water dish was filled and the appliance allowed to run for 24 hours, it was ready for use.

The first eggs I tried were 14 Maran eggs. Nine of them hatched successfully. If I were constructing it again today, one of the refinements I would include would be an electronic thermostat, for there is no doubt that it is in adequate temperature control that most DIY incubators are lacking. By comparison with most bought incubators, the hatching rates I achieved were poor. Using my normal bought incubator, the success rates were far higher.

Some time after this design appeared in *Home Farm* magazine, I heard from a reader who had decided to build it. This is what he had to say.

Barry Gillett's comments

"The dimensions I used were slightly different from yours but the idea is the same. I had two 40 watt bulbs controlled by an immersion heater thermostat of the type used in 'over-the-sink' heaters (*Satchwell VK1 1201*) cutting in at 96°F and out at 104°F. The temperature and humidity are monitored by a pair of gardener's gauges made by *West Meters* and costing under £2 the pair. Temperature was no problem at all; the thermostat saw to that. Humidity was a different matter; I could find no way to control it with any degree of accuracy.

I tested the incubator for 24 hours; it worked well and quail eggs were installed. They were turned five times a day. At each turning warm water was added to the reservoir, so that humidity was maintained. Last thing at night and first thing in the morning the eggs were sprayed with warm water. The first batch was set for 21 days with no result. Having read 16-19 days was normal, I switched the incubator off. A post-mortem revealed that 4 eggs contained perfect specimens, one had formed but not grown, and the remainder appeared infertile. Another day might have seen a hatch, if only I had been patient.

I was very angry with myself; it was a very hard lesson. Off we went and bought two hens and some fertile eggs. When we got them home we examined the eggs very closely and three were discarded because of hairline cracks. We had nine left, the original number. The incubator was turned on, the eggs installed and allowed to warm up with the incubator. Religiously the eggs were turned five times each day, warm spray night and morning and humidity kept between 60-70%. On day 14 turning was stopped, humidity and temperature were maintained and on day 18 they began to hatch. The children, June and myself watched as each tiny inmate

forced its way out into its new world. As each egg hatched the excitement mounted. Six times we saw it happen, six times we were filled with wonder. Our daughter, her face a mixture of surprise, amazement and curiosity, encouraged each chick from its shell. The failure of the first batch made this one very special; it was the proof that the incubator really worked; doubts had filled my mind right up until the first egg pipped. We have all learned something from this experience. We would not, I feel sure, have gained as much pleasure had we bought an incubator. Thanks to *Home Farm* for the idea and the pleasure you have given us".
(Barry Gillett)

Carl Garnham's home-made thermostat incubator

Carl Garnham is a breeder of pheasants, quail and water-fowl. This article, where he describes his home-made thermostat incubator, was also published in *Home Farm* magazine. It should be pointed out that only those with a good knowledge of electricity should attempt to adapt the thermostat as he describes.

"While the incubator I shall describe may sound extremely crude and unreliable, I can assure you, from personal experience, that this is far from the case. The design has hatched countless pheasants, quail, ducks, geese and, from personal experience again, I can guarantee that it will control both temperature and humidity to a good tolerance. The incubator can be constructed of any wooden or even thick, rigid plastic sheeting, but probably by far the cheapest is half inch thick exterior grade ply, particularly shuttering grade. Off-cuts can often be scrounged from building or construction sites. (Despite its initial appearance, once well sanded and varnished, it can look exceedingly smart.)

Exact dimensions should depend solely on one factor - the size of water tray that can be obtained. The incubator should measure internally up to approximately twice the area of the tray, eg, a tray measuring 8" x 5" could be used in an incubator with a floor area of up to 80 square inches - flower pot saucers and photographic processing trays. Vertical dimensions should be of the order as indicated - Diag 1, page 56.

Insulation, though not essential, will reduce the already meagre running costs and also tend to allow a more steady control of temperature. This is effected by using sheets of expanded plastic foam. Both polystyrene and polyurethane are used for domestic insulation and hence off-cuts, again, can often be begged from building sites. If not, sheets measuring 4' x 2' can be bought cheaply from a good builder's merchant.

Heating of the incubator is via two or more 15 watt pygmy bulbs. Many colours are available but make sure that all in use at the same time are of the same colour as different colours radiate heat at different rates, and hence lead to uneven heat distribution in the incubator. These are wired in 'parallel', see Diag. 2, so that should

Carl Garnham's home-made thermostat incubator

1.

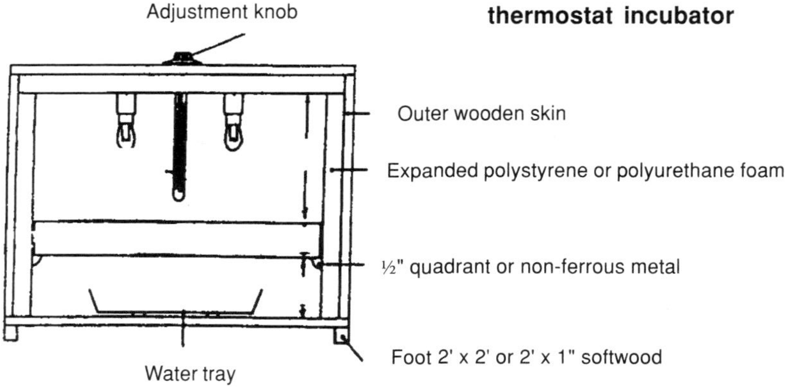

2.

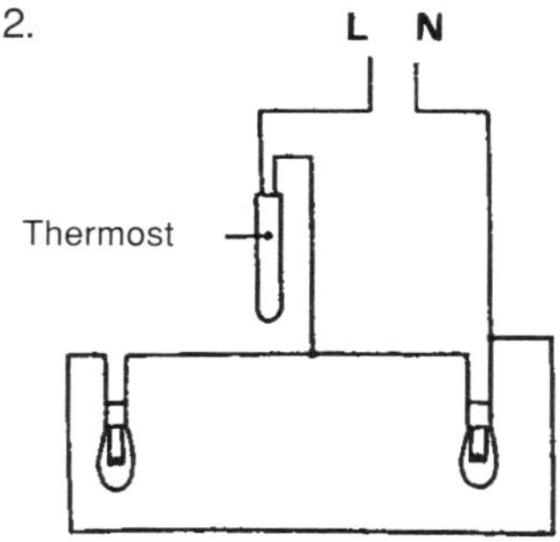

Parts list

Shuttering ply 8' x 4' sheet
Insulating foam sheet 4' x 2'
2 x Pygmy bulbs
Water tray
Half inch hardwood quadrant
2 x Bulb holders
1 x Adjustment knob
1 x Fish tank thermostat
Miscellaneous screws and nails

3.

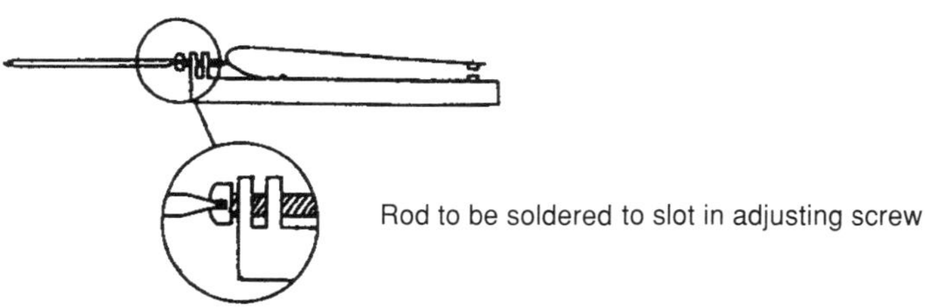

4.

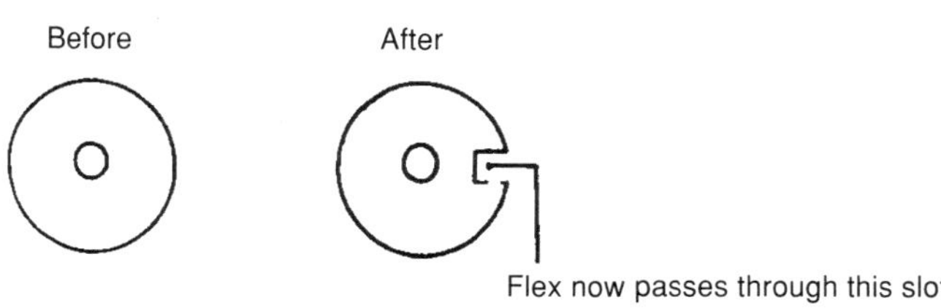

one blow, then the other will continue to operate unhindered. The bulbs have a life span of approximately one and a half full breeding seasons and should hence be discarded after one. In the event of one of the bulbs blowing, although the temperature in that part of the incubator may drop appreciably, I have never found this to be a problem as the remaining bulb keeps things quite warm and the fault should be detected after no more than 6-8 hours on turning the eggs.

The electrical circuit should be fitted with a one amp fuse. If this continually blows then a mistake has been made in the wiring. In my experience, two bulbs are ideal for heating an incubator measuring approximately 15" x 8".

Despite much information to the contrary, I have found the crude bi-metallic strip type fish-tank thermostat to be capable of giving entirely adequate control of temperature. However, there are two main problems encountered with this type of thermostat. Firstly, they are slow acting, ie, they control over a temperature range rather than at a precise temperature, so that a source of heat that very quickly radiates the necessary heat is essential, and, secondly, unless modified, their adjustment entails dismantling the instrument. To allow external adjustment is simple - see Diag. 3. A short length of $^1/_8$" diameter metal rod (welding or brazing rod is ideal, but not aluminium) is flattened at one end and then filed to give a clean surface which tightly fits the slot in the head of the thermostat adjusting screw. This is then soldered into place in the slot. To allow the passage of both adjusting rod and connecting wire from the tube, the rubber bung is cut as in Diag. 4.

As the metal adjusting rod will become 'live' on connection, the rod should be covered with wooden, or better, plastic, rod, which should preferably be of the standard ¼" diameter to allow the attachment of a knob, readily available from any electronics hobbyist shop. This sheath need only extend down the rod as far as the surface of the incubator and should be rigidly fixed with a strong glue.

In all probability it will be found that if attempts are made to adjust the thermostat's set-point, the body of the instrument, ie, the bi-metallic strip and the piece of plastic to which it is attached, will merely rotate within the tube. To avoid this, a thin narrow strip of cardboard should be glued to each side of the plastic part of the body so that the body will become a tight push-fit in the tube and hence unlikely to move. All that remains is to remove the rubbery plastic coating from the thermostat tube as this is a comparatively good insulator with respect to heat and hence 'dulls' the accuracy of the instrument because of this. (If it is found that incubating temperature is beyond the range of the thermostat as received, a small amount of judicious bending of the bi-metallic strip itself will quickly cure it).

The egg tray, whether the incubator is to be used as a hatcher or not, must not allow the eggs to move except on turning. I have found ½" - 1" Twilweld absolutely ideal for this, but if used for a hatcher it must be under-wired with smaller mesh as this is really too large for young birds to stand on. The tray is supported on either small strips of non-ferrous metal or lengths of hardwood quadrant.

Ventilation is dictated largely by humidity. Felts (pieces of hessian sacking or carpet underfelt are adequate) can be added beneath the water tray to reduce the through flow of air and consequently increase relative humidity. The necessary ventilation holes should be of around $^3/_8$" in diameter and number no more than six in both top and bottom. In conclusion, use as little ferrous metal internally as possible (warm, humid conditions are ideal for encouraging rust) and shop around for materials.

Temperatures maintained by the thermostat will be steady to within well under half of one degree Centigrade and quite possibly less than one fifth of a degree. However, the bi-metallic strip reacts to the temperature along its length, which by the nature of the method of heating, is not constant; it is a temperature gradient. As such the temperature at which the thermostat will control will, without adjustment, vary as the eggs develop and start to evolve heat themselves, hence changing the temperature gradient. Minor adjustment should be needed approximately every other day." (Carl Garnham)

Carol Alvin's incubator

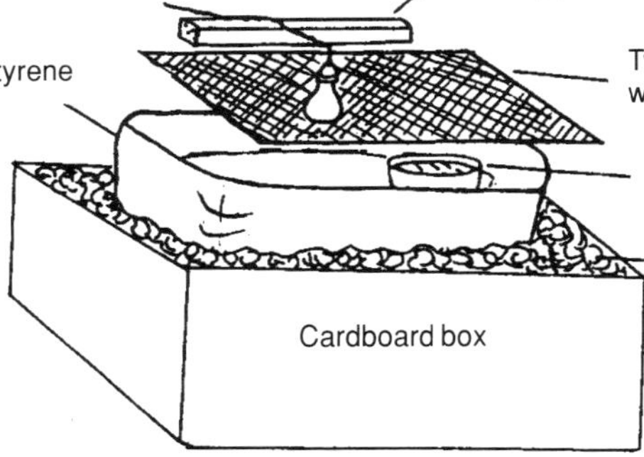

This was an idea from another reader, Carol Alvin, whose suggestion was also published in 'Home Farm'. Her incubator has no thermostat, so is liable to temperature fluctuations, but she obviously has had success with it.

"I used three expanded polystyrene boxes 16" x 10" x 5". Two had their bottoms cut out and were taped on the third to give a box 16" x 10" x 15". There is no need to be fussy with the joins as you will need some air circulating. If you use a box of different size or material you will have to experiment with bulb wattage and air holes to get the temperature correct. I found that with a single 40 watt bulb, I needed to line the two short sides of the box and the lid - a piece of hardboard from a tomato box - with aluminium foil to reflect some of the heat. The bulb was fitted into a lamp holder fixed to a rod across the top of the box. My holder happened to have a grip attachment but string would do just as well. (I have also tried a small table lamp minus the shade in a larger box with some success).

The whole contraption was then put into a cardboard box about 2" larger all round, and lightly crumpled newspaper packed into the space. Don't site the incubator in a draught, nor in a room where there are marked fluctuations between day and night-time temperatures or you will be for ever adjusting the set-up. I chose the bottom of the wardrobe (ours has a curtain in place of the usual door) in the bedroom as offering the most even temperature, and adequate ventilation.

Muscovy duck with her brood in a rabbit hutch from which they can range at will.

In the bottom of the incubator you will need a dish of water and some means of preventing the eggs rolling around. I solved the latter problem by cutting a piece of rigid plastic mesh (found on the beach) to fit inside the box and balanced it on top of the water dish which could be topped up by pouring water through the mesh. The eggs were arranged on the mesh and thin pieces of wood could be inserted through the mesh, pegboard fashion, to hold the eggs in place.

Having set it all up, the next step is to switch on the bulb and run the incubator without eggs until you can maintain a steady temperature of 39°C. I use a dairy thermometer because it is all I happen to have and it is also very easy to read. You will need to play around with the lid position, more or less insulation, and/or air holes until the temperature remains steady for several hours. Also check, by moving the thermometer around, the temperature of different parts of the box. I put a false egg directly under the bulb as this position tends to be too hot when the rest of the box is just right.

When the temperature has stabilised, put in your eggs. Use the freshest eggs you can get. From experience I have found that eggs over ten days old will not hatch by this method. Perhaps they are more susceptible to the inevitable fluctuations of the temperature that occur. Put the lid on tightly and leave for two or three hours to come back up to the correct temperature. Thereafter it is a question of checking the thermometer, keeping the water topped up, and turning the eggs every four or five hours for the first fortnight, then every six hours until day nineteen.

Adjust the lid as necessary. If you are lucky you will get a three week spell of stable weather and little adjustment will be needed, but if there is a dramatic change outdoors be sure to check that thermometer! By the 19th day stop turning the eggs. You can find out how many live chicks you have by placing the eggs briefly in a bowl of warm water. The live ones will bob around quite vigorously". (Carol Alvin).

Ostrich chicks in brooding conditions. Note the rubber mats to prevent slipping - *Maywick*

Part 2
Rearing

Make arrangements before you need them!

The brooder

Once the chicks are dry and showing an interest in their surroundings they are ready to be moved from the hatcher to a brooder where temperature levels can be maintained at a relatively high level, but where food, water and exercise are available. Brooding is the period after hatching when young birds must have protected conditions until they have reached a full-feathered, more hardy stage of their lives. It is at this stage that they are at their most vulnerable to temperature fluctuations, chills and so on, and particularly to the onslaught of rats.

Make sure everything is ready in good time, including cleaning and disinfecting the brooding area, feeders and drinkers, and checking the heat lamp. Ensure that the flooring is secure, clean and well littered with several inches of new poultry wood shavings. They are available in compressed bales from numerous sources.

A shed, outbuilding or hen house can be adapted for brooding purposes, as long as it is totally rat-proof. Concrete floors are the most effective in this respect, for rats are highly intelligent creatures devious in finding access to their prey. Wooden floors may need some weldmesh to make them secure.

The brooding times will vary, depending on breeds and type of birds. Outside conditions will also have an effect. It is easier to brood young birds in the warm summer months than it is in winter. There is some evidence to indicate that chicks which are able to get outside at an early stage will feather more rapidly, and are generally more hardy. If outside conditions are suitable, the developing chicks can be allowed access to an outside run, returning to their indoor quarters for the night.

Brooder lamp

A radiant heat brooder lamp in a protected area inside is the most common way of brooding chicks of all kinds The type of bulb is important because if too much bright light, rather than heat is emitted, not only is this distressing for the chicks, but it may also encourage toe-pecking and other forms of bullying. Red bulbs are sometimes used where this is is a problem.

Electric brooder lamps are widely available. One 250 watt lamp will cater for 80 chicks. An alternative, where scale permits, is to use gas-powered ones. Paraffin brooders are still around although they are not as common as they once were. Hot water brooders were also used at one time, but again, they are no longer common.

General temperature needs	
1 day old	35°C
1 week	33°C
2 weeks	30°C
3 weeks	28°C
4 weeks	25°C

Brooding unit in an outbuilding with corrugated cardboard used as a brooding ring.

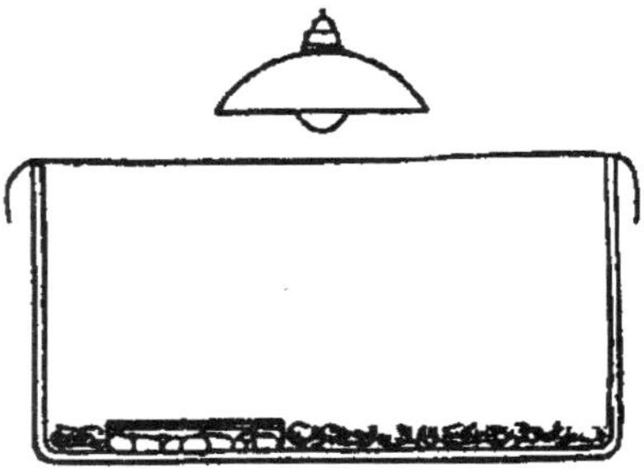

A safe way of brooding tiny quail without losing them is to utilise an old fish tank with netting across the top.

If chicks cluster in a ball under the lamp, they are too cold - lower the lamp. If they are ranged around the periphery they are too hot - raise the lamp.

A rabbit hutch adapted as a brooder.

A large cardboard box with an *Anglepoise* lamp can also be used as a temporary brooder for a small number of chicks.

Adapting a poultry house as a brooding area. Take care to exclude rats.

The correct height of the lamp is easily established: if the chicks huddle under the lamp in a tight mass, they are cold and it needs to be lowered. If they are hovering around the edge of the heat source, it is too hot, and the lamp should be raised. As they grow the lamp is gradually raised in order to harden them off, but the first three days are crucial and they must be kept warm to prevent chilling.

Some kind of partitioning will be required to confine the chicks to the area where the lamp is situated, and to prevent them escaping when the door is open. Some manufacturers supply portable sectioned *brooding rings* to go with a suspended lamp. An alternative is to use corrugated cardboard placed in a circle around the lamp. As the chicks grow and range more, the walls can first be extended then discarded altogether. They are generally only necessary for the first week.

There are some purpose-built *canopy* or *hover* brooders which have a 'skirt' around the bottom. They emulate the mother hen in that the chicks are able to dive under the 'skirt' for protection. They are available in a number of shapes and sizes.

Flooring

A thick layer of poultry wood shavings makes a warm, cosy floor. As concrete is a cold material, several layers of newspaper can be laid down first before applying the wood shavings. Newspaper on its own is slippery and can cause leg damage.

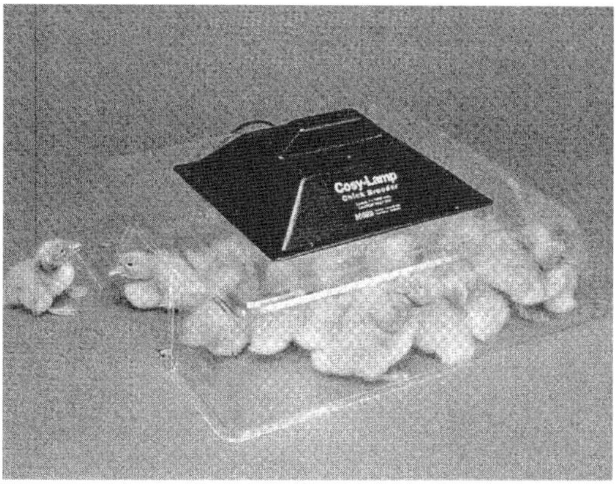

Top: The *Cosy-Lamp* brooder suitable for a range of different chicks. *Brinsea.*

Middle: The *Ecostat* brooder with a 'skirt' to emulate the mother hen's protection. *Ecostat.*

Bottom: A *Brinsea Octagon* incubator with alternative cover in use as a brooder for a parrot chick. *Brinsea.*

Chicks of large birds such as ostriches are best provided with ribbed rubber mats on the floor. These enable them to get a grip and prevent slipping. They also do away with a situation where the ostrich chicks might eat the litter and suffer digestive impaction, a condition to which they are particularly prone. Smaller birds can be trained not to eat litter by covering the wood shavings with paper towels when they are first put into the brooder. Food can then be placed low on the ground, and if necessary sprinkled to draw their attention to it. Once they are eating the feed, the towels can be removed.

Maintaining the litter in a dry, clean condition is obviously important in the avoidance of disease. Coccidiosis organisms thrive in damp conditions and the developing birds can also develop breast blisters from damp litter with a high ammonia content. Damp litter also turns mouldy, providing ideal conditions for the development of *Aspergillosis*.

Some brooders utilise under-floor heating. There needs to be a thick protective layer above the heat source that prevents the floor becoming too hot. There should also be a thick layer of clean litter on top of that. If the rising heat is not properly controlled, it can damage the chicks' feet.

Temperature

When first put in the brooder, the chicks will have come from a hatcher that was around 35^{0}C. It is important that they are not suddenly plunged into a much colder environment. As a general rule, this temperature should be maintained for the first few days, gradually decreasing by 2-3 degrees each week until it is around 25^{0}C by the fourth week. Gradually raising the heat lamp lowers the temperature.

Outside weather conditions will also have a bearing. If it is really fine, chicks should be encouraged to go outside, not only because it is an incentive for rapid feathering, but also because the exercise is good for their legs. The area to which they are introduced should be one that has not been used for poultry in the same season, so that the possibility of disease and parasitical transference is minimised.

Feeding and drinking

A feeder for chick crumbs and a drinker for water are the only other requirements in the brooder. A suspended drinker is preferable to a floor-standing one because there is less likelihood of droppings getting into the water. Both feeders and drinkers should be carefully placed to get the height right. They should be at a level where the chicks can gain easy access without having to strain.

Chicks without a mother to show them how to eat may take some time to work it out. If there are some slow ones, imitate the mother hen by picking up some of the chick crumbs then drop them onto a sheet of paper on the ground. The sound and action stimulates tardy chicks into investigating. Similarly, pick them up gently and dip their beaks in water; they'll soon learn to drink!

Reference has been made to chick crumbs which provide the best start for young

Turkey poults under a *Maywick* gas brooder. They are housed in an outbuilding and confined to the warm area by a brooder ring. Note how some feed is placed in shallow dishes for increased accessibility until they have learnt how to feed. *Maywick.*

birds. They are also referred to as a *starter ration* and are high in protein. They are available for specific birds and it is advisable to buy the one that is appropriate, eg, turkey starter for turkey poults, etc. These should be made available on an ad-lib basis so that the chicks can help themselves when they feel like it.

Some, such as chick crumbs for young chickens, contain coccidiostats, antibiotic additives to counteract the disease coccidiosis, but there are specialist feed suppliers who formulate chick crumbs without them. Those who are not raising large numbers of birds intensively prefer to use these more natural feeds. Organic producers also use them. It is worth remembering however, that there is a greater risk of coccidiosis if the chicks are on the ground outside. Even where runs are placed on areas where poultry have not been ranging, coccidiosis can still be introduced by wild birds. Many breeders compromise by letting chicks have a starter ration containing a coccidiostat while they are at the vulnerable young stage, then switch to a natural feed without additives later. Turkey starter rations, and those for game birds, often contain an additive against the protozoan Blackhead (*Histomoniasis*). Waterfowl are not usually given additives in their rations, and a waterfowl starter ration should be given. Geese, in particular, may react to some additives.

Traditionally, finely chopped hard boiled eggs were fed to chicks, along with finely ground cereals, and this was undoubtedly a good, high protein start, although labour-intensive in its application. A small amount of natural, live yoghurt is also recommended by many, not only as a suppressant of *Salmonella enteriditis*, but also to help the natural flora of the gut to develop. It acts as a balance against the coccidiostat in chick crumbs which disrupts the natural bacteria of the gut while it is killing off pathogenic bacteria.

Those new to poultrykeeping may be concerned when incubator hatched chicks appear not to be eating. It should be remembered that the remains of the yolk are still in their abdomens, and they are adequately fed for the first two days after emerging from the shell. (This is why it is possible to transport day old chicks without needing to worry about feeding them en route, and the new transport of livestock legislation reflects this). If there are still a few slow learners, teach them how to feed and drink, as detailed earlier.

Once the chicks are 5 weeks old they can be switched to a lower protein *grower's ration*. These are available for a range of different birds. Again, it is possible to get them without additives. Grain is also required at this time, as well as insoluble grit for its proper digestion. Some breeders prefer to give grain at an earlier stage. If this is done, it is important to provide *kibbled grain*. This is grain that has been chopped so that the smaller pieces are more appropriate for chicks.

Vaccination

Commercially chicks are always vaccinated, unless there is a specific requirement such as table birds being reared organically. On a small scale, should one vaccinate chicks? The question is an individual one and it is true to say that many small poultry keepers do not do so, but there are three situations where it is essential to vaccinate against Marek's disease, Newcastle disease and Infectious bronchitis.

〈 where the birds are to be sold - in order to protect the buyer from potential problems and the seller gainst possible redress.
〈 where the birds are future breeding stock.
〈 if a locality has a prevalence of these diseases. A veterinary surgeon will advise.

If it is decided to go ahead with vaccination, these are the recommended times:

Marek's disease	day old
Newcastle disease (Fowl pest)	2 weeks
Infectious bronchitis	4 weeks

Vaccines are also available for Gumboro (Infectious Bursal disease), Infectious laryngotracheitis, Epidemic tremors (Avian encephalomyelitis), Egg drop syndrome, Fowl pox and Fowl cholera (Pasteurellosis). As far as the small breeder is con-

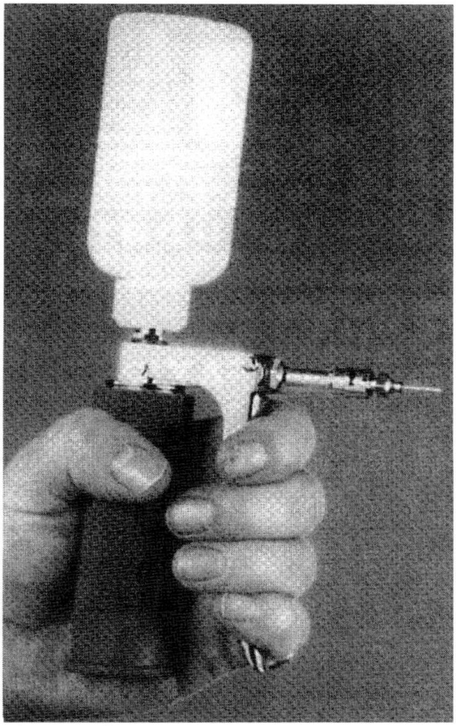

Kaycee Supervax 5000, an automatic syringe for vaccinating chicks.

cerned, no action need generally be taken against these unless they appear in neighbouring flocks, and the vet recommends action. It is obvious that anyone who is going in for breeding will liase quite closely with a veterinary surgeon.

Vaccines are available in one of two forms - *live* which are made available in the drinking water, and *deactivated* which are administered by injection. The advantages of the latter are that they are usually available as *combination vaccines* so that several diseases are covered at the same time. They can also be administered with an automatic or multi-dose syringe so that each bird receives an accurate, calibrated dose. The procedure is one that needs to be demonstrated however, and a poultry husbandry course at one of the agricultural colleges is advised. Accidental self-injection is a hazard which must be dealt with as a medical emergency, so the use of deactivated vaccines is not recommended to the untrained user. Vaccines are only available in large quantities usually 1,000 doses), but the cost is relatively low (around £2-£3 for a 1,000 dose vial of Infectious bronchitis vaccine). There is a large wastage as far as the small breeder is concerned, but the manufacturers state that it is not cost-effective to package vaccines in smaller quantities. Live vaccine vials are easily diluted to the required dilution by following the manufacturer's instructions.

The broody hen

The broody hen is a most efficient incubator, turning the eggs about once every half an hour and keeping them warm by radiating heat, particularly from her breast area. The trouble is that a broody hen may not necessarily be available when eggs are waiting to be incubated. In that case, having an incubator is a positive advantage. It is often claimed that traditional breeds are more likely to become broody than hybrids, because the tendency towards broodiness has been bred out of hybrids. The Ross Brown on the outside front cover of this book is a second-year commercial hybrid, yet is the epitome of the traditional broody.

Some bantams do appear to have a greater trait for broodiness than many larger birds, and most poultry keepers will know of a banty who only needs to find a couple of discarded door knobs to become broody. Such birds are a boon to the

One of the author's bantams with the chicks she has hatched. They are in a small house and run to protect them against predators.

small breeder, but their small size does limit the number of eggs they can cover. Because of this a popular cross to produce a reliable broody, but with a substantial size, is Silkie x Light Sussex. A broody hen will generally find her own place to make a nest, but, for her own protection, it is better if she can be persuaded to use a purpose-made broody house. She will usually leave her nest once a day to feed, drink and relieve herself. Food and water should be available close at hand and if there is no confining run, she will usually go quite a distance away to deposit her droppings - a tendency which is a survival mechanism dating back to the wild.

While confined in a warm environment such as a broody house, she is prone to lice and mite attack. It is important, therefore, to give her a good dusting of proprietary powder, available from licensed suppliers, *before* she starts to sit.

To introduce fertile eggs to a sitting broody, carefully remove an egg at a time and slip a fertile egg under her. Let her accept one before introducing the next. Remember to mark the fertile eggs otherwise you may get them mixed up.

To introduce newly hatched chicks from an incubator, the trick is to make her believe that the eggs she has been sitting on for several weeks, have hatched. Slip one under her, preferably from the side, or from behind, and see if she settles herself anew over it. Introduce more chicks, removing enough eggs to make room for them. Watch her carefully to ensure that she has accepted the chicks. If she has, she will cluck to them in answer to their cheeps, and will seek to cover them protectively.

Home-made brooders

Ingenuity is a useful characteristic, particularly where special needs must be catered for. Ouail chicks are so small that it can be a problem to cater for them effectively. The diagrams on page 62 show some suggestions. The following are some reader ideas contributed to *Home Farm* (now *Country Garden & Smallholding).*

Carl Garnham's brooding unit

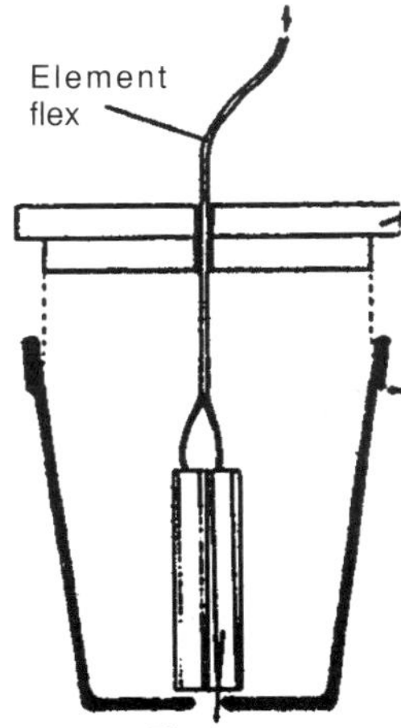

"My system comprises a set of three elements, rated at 25, 16 and 8 watts. These are generally used inside an ordinary 6" clay (terracotta) flower-pot that is placed on the floor of the brooder - the pot acting as a surrogate mother in that the young birds will sit against the pot for warmth as they would their mother. The set of three elements enables the birds to be 'hardened-off' in stages while, at the same time, saving money due to the reduced power consumption.

While for the vast majority of birds that I rear, 25 watt is entirely adequate to start with, if the birds are to be reared in either very cold weather or generally cool surroundings, a draughty barn for instance, 37 watt elements are also available. Dependent on the ambient temperature of the surroundings in which the brooder is situated, a single pot will brood at least 12 quail, 8 small ducklings or pheasant chicks or 4 goslings or cygnets. If used for goslings, cygnets or large ducklings, however, the pot must be weighted with dry sand, pea-gravel or similar, to prevent the birds from knocking over the pot. The element should then be buried in this. In all cases, it is desirable though not essential to use this method as the sand/gravel produces an even heat around the pot.

In extreme circumstances more than one element can be used in one pot or two pots used close together.

To avoid unpredictable failure of the heat source, each element is made as at least two windings. In addition, as the elements operate at comparatively low temperatures, the life expectancy of the elements is considerable, given that they are not subjected to unreasonable treatment. It is wise to check the elements daily by merely using your hand on the outside of the pot - you will quickly get to know the 'correct feel'. To prove warm enough to successfully brood newly hatched youngsters, the outside of the pot should reach a temperature of at least 45°C (the pot may take up to 6 hours to reach maximum temperature, especially if the pot is damp). Depending on conditions, the following is a reasonable schedule to work to:

Ducklings: 25 watt - 0 - 7 days old, 16 watt — 8 - 14 days old, 8 watt -15 - 21 days old.
Ouail: 25 watt — 0 - 10 days old, 16 watt — 11 - 20 days old, 8 watt -21 - 30 days old.

If conditions warrant the further use of heat, 7 watt can be used at night only for some days after this. Goslings and cygnets generally require less brooding, periods of approximately 4 to 5 days at each brooder temperature often proving adequate. In all cases, if the young birds appear cold after changing a setting, give them an extra couple of days on the higher setting.

The elements I produce are merely sophisticated light bulbs - a method long used in this area for brooding small numbers of birds. Be that as it may, I am constantly surprised by people who claim never to have seen or thought of using an ordinary light bulb for brooding, either on its own or inside a flower pot: 40 watt in cold weather or draughty situation, 25 watt in better conditions and hardening-off by use of 15 watt bulbs and 8 watt night-lights. The disadvantage is the unpredictable life of bulbs so that unless more than one is used with each batch of birds, disasters can be unpleasantly common."
(Carl Garnham)

Carol Alvin's brooder

"I transfer chicks to a cardboard box with an enclosed bulb and an inverted 'mop' of strips of woolly fabric suspended a few inches above the floor of the box. They have access to an unheated area and by their second day they are running back and forth, and eating voraciously. Given reasonable weather, I gradually wean them off heat and they go outside into a well strawed hutch at five weeks.

As for the mop, it was a tip given me by my father who used to help out at a large poultry farm in his youth 50 years ago. It is supposed to simulate the hen and offer a little extra comfort that a harsh light doesn't do however warm. I use strips of old jumpers, about 4 cm wide, tied together in the middle and suspended from the same rod as the bulb but on the other side of the box.

The mop doesn't touch the bulb at all so I don't think it would be any more of a fire hazard than the cardboard box itself. The chicks seem to appreciate it and can often be found all together in a heap under it. At this stage I use a bulb enclosed in a wire inspection holder - this keeps the chicks from burning themselves on the bulb."
(Carol Alvin)

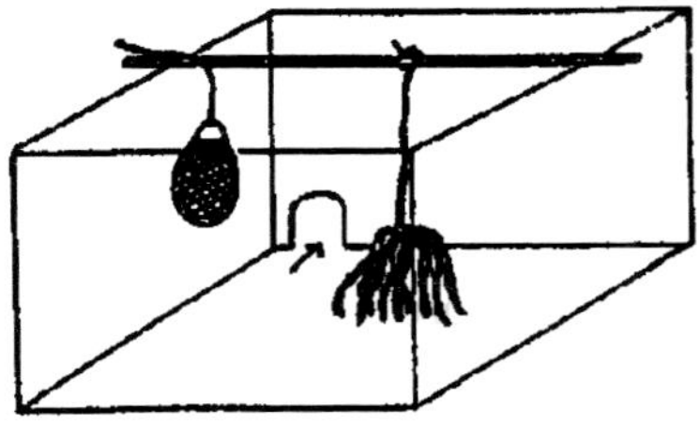

Carol Alvin's brooder . The arrow shows the way through to an unheated area. The suspended strips emulate the mother hen

Chick assessment

The chicks have hatched successfully but it is important to keep track, not only of their development, but of their origins if they are part of a selective breeding programme.

Recording

As soon as the chicks are removed from the hatcher, record results! This will normally entail details of how many eggs were started, how many hatched, the number of infertile, dead-in-shell, and so on. Other relevant information will be how many chicks hatched from a particular mating, the parents' identification, the sex, etc. Some suggested record cards are given below, but everyone will have their own preferences, so they can be adapted as necessary. It is also a good idea to have *Brooder Records*, *Parent Records*, and indeed any other you think fit.

MONTHLY TRAP NEST RECORD

Month __________ **Breed** ______________ **Ring No** ____

							Weekly Total

Monthly Total

Incubator Record

Date set _______ **Date due** ______

Breed ____________________

Parent Breeders ________________________________

Leg Ring Numbers ______________ ______________

No of eggs set _____ **No of infertiles** ______

No of chicks hatched _____

No of chicks dead in shell _____

No of chicks removed to brooder _____

Notes:

(See page 94 on how to work out the average fertility and hatching percentages).

Identification

As soon as possible, the chicks should be identified, if they are part of a selective breeding programme. This may involve the use of wing tags or toe-punching. The latter is where a small section of the web between the toes is cut. Depending on whether the right or left foot (or both) are marked, and in between which toes the knick occurs, there are many combinations. Any of these can be used and recorded in order to identify young birds. When they are older, they can be fitted with identifying leg rings. Large, expensive birds such as ostriches and emus are often electronically tagged

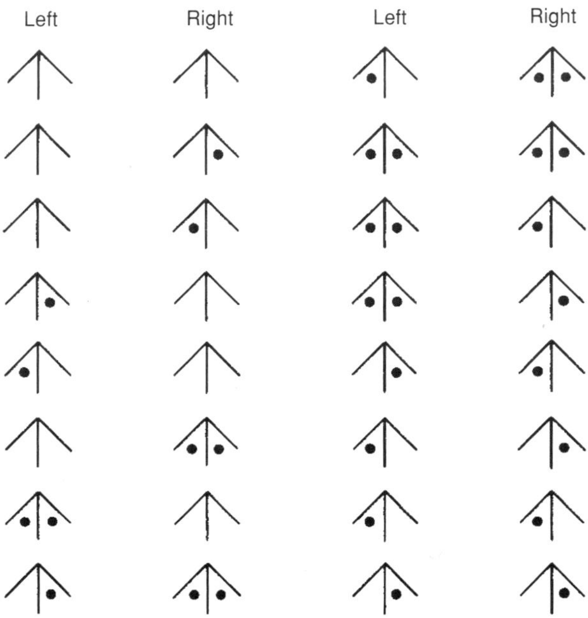

The possible combinations of toe marking for identification of chicks

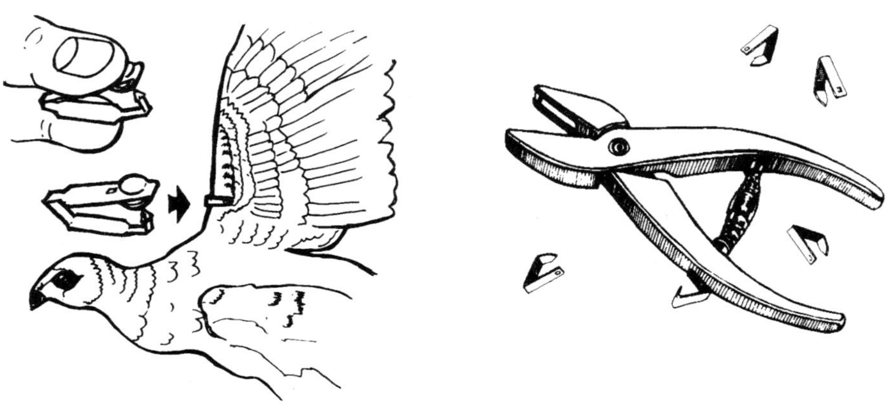

Wing tagging young birds - in this case pheasants. *Patrick Pinker (Game Farm) Ltd*

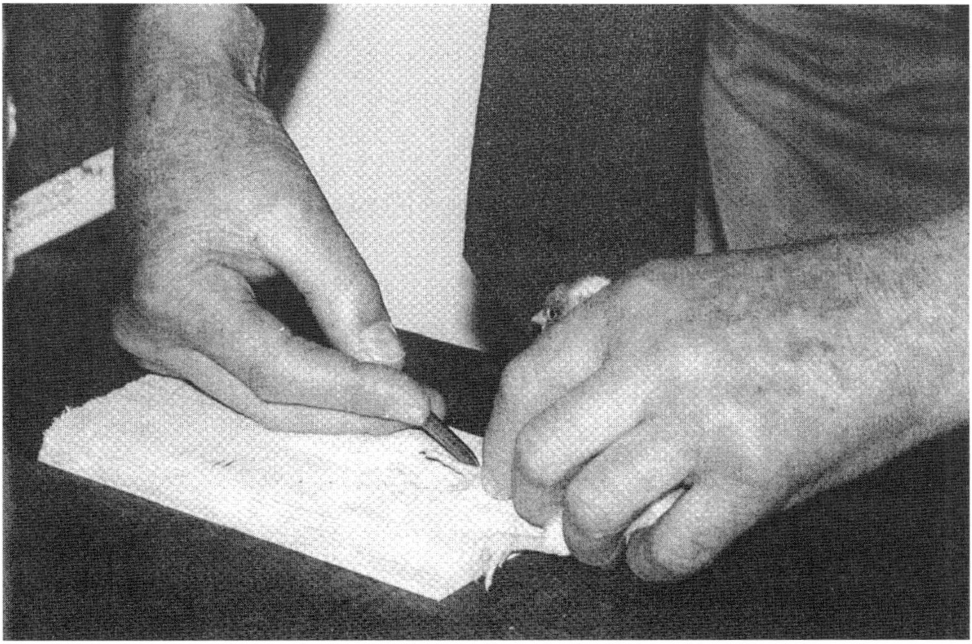

Toe marking newly hatched chicks. In this case a sterilized scalpel is being used to make a small nick in the web in between the toes. It is a painless procedure.

Sexing

Being able to tell which young birds are males and which are females is often a problem, unless a particular feature such as colour variation is apparent. It is a skill which, professionally, takes a lot of experience to master. It is also a field where popular mythology abounds! The following is a round-up of both.

Sex-linkage

The easiest way to tell the difference is where certain crosses are sex-linked.ln other words, the male chicks show one colour or patterning, while the females show another. The most well known of these is the Rhode Island Red male crossed with a Light Sussex female. The males will be silvery yellow, while the females are buff-orange. This particular feature is apparent where any 'gold' male is crossed with a 'silver' female, as the list overleaf indicates.

Other crosses such as using 'black' males on 'barred' females will produce easily distinguishable chicks, for the males will be black with a light patch on the head, and possibly the wings, while the females will be all black.

Dark-legged males crossed with light-legged females tend to produce young which demonstrate the opposite characteristic. In other words, the males will be light legged, and the females dark-legged.

Light and dark eyes is also a feature which can be used with some breeds, where a dark-eyed father and a light-eyed mother produce dark-eyed female chicks and light-eyed males. It is important to remember that these sex-linked features do not work the other way round, where the breeding pair is reversed. Nor will a second cross necessarily show the distinctions. Bear in mind that some birds which are claimed to be pure-bred may not be. They may be a good phenotype in that they look like the appropriate breed, but the genotype may include some genes that are not apparent on the surface. Nevertheless, as a general rule of thumb, you can still use the features as a guideline.The important thing is that you do not rely on the accuracy otherwise you may be disappointed.

Mating any of these 'gold' males with any of these 'silver' females will produce male chicks which are light coloured and females which are buff coloured.

Males	**Females**
Barnevelder	Light Brahma
Brown Leghorn	Light Sussex
Brown Sussex	Silver Grey Dorking
Indian Game	White Wyandotte
Partridge Cochin	Ancona
Partridge Wyandotte	
Buff Leghorn	
Buff Rock	
Buff Orpington	
Rhode Island Red	

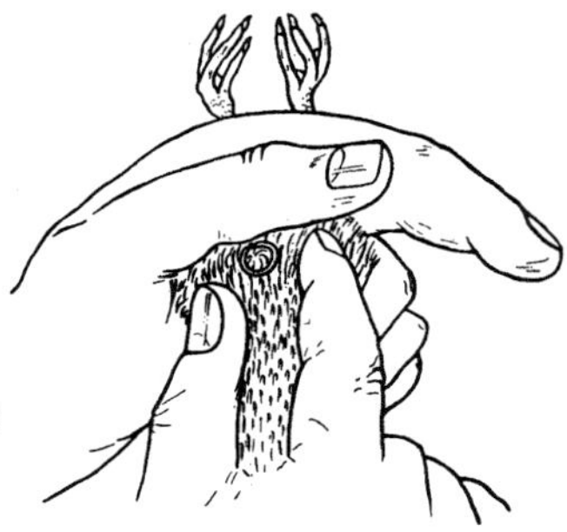

Right: Method of holding chick for vent examination

Cross any of these 'black' males with any of these 'barred' females, and the male chicks will be black with a light patch on the head and possibly also on the rump, while the females will be all black. (They may also have a head spot, but it may be smaller than the male's - if you are lucky!)

Males	**Females**
Australorp	Cuckoo Leghorn
Croad Langshan	Cuckoo Dumpy
Black Leghorn	Barred Rock
Black Langshan	
Black Minorca	
Black Orpington	
Black Silkie	
Spanish	
Black Wyandotte	
Black Cochin	

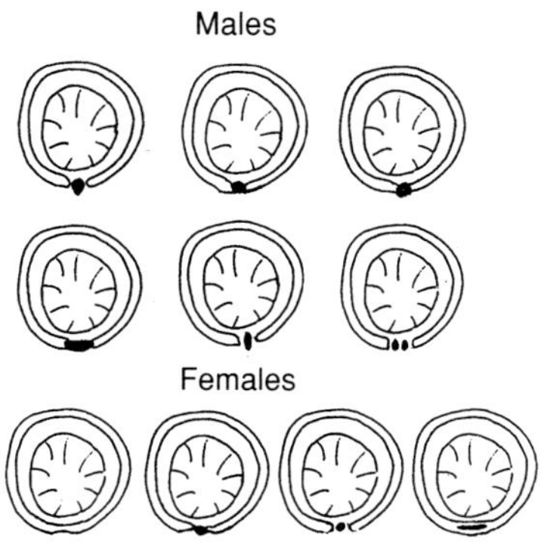

Right: Some of the possible vent variations

Crossing the following 'dark-legged' males with any of the 'light-legged' females will tend to produce males with light legs, while the females are dark-legged.

Males	**Females**
Australorp	Barred Rock
Black Minorca	Buff Rock
Brown Leghorn	Brown Leghorn
White Bresse	Buff Leghorn
Buttercup	Light Sussex
Campine	Scots Grey
Hamburgh	White Wyandotte
Langshan	

Again, a word of warning!
These combinations cannot be taken as fool-proof method of sexing chicks, for many birds that are claimed to be pure-bred are not necessarily so.

Cross a dark-eyed male with a light-eyed female, and the chicks will tend to be dark-eyed females and light-eyed males.

Male	**Female**
Langshan	Brown Leghorn

Autosexing

Sex-linked crosses were popular with poultry breeders, but required the keeping of two separate breeds, and first-crosses could not, in turn, be used for breeding with the same certainty. Then, auto-sexing breeds were developed, which would show sex differences at day-old and yet would still breed true. The first breed to meet these requirements was the Gold Cambar, shown at Crystal Palace in 1930. This was originally produced from the Gold Campine and Barred Rock, bred over several generations. After that, research produced the Legbar from a Leghorn Cross, a Dorbar from a Dorking Cross, and a Welbar from a Welsummer Cross. All the autosexing breeds were developed by adding the factor for barring into an unbarred breed. It is sex-linked, behaving in a similar way as the factor for silver, and males have a double dose while females have a single dose.

Vent-sexing

Vent-sexing is a most difficult method for the amateur, and the consensus view seems to be that a professional sexer requires at least a year's experience before a degree of accuracy is achieved. The technique appears simple enough, but appearances can be deceptive. Pick up the chick as shown in the illustration, and apply slight pressure in order to make the vent protrude slightly. (Great care is needed to avoid causing physical damage to the chick).

The problem now is in interpreting what is seen, for there can be a considerable variation. In fact, it has been claimed in a treatise published by the *Chick Sexing Association* of Japan, that there are over sixty variations in the sex formation of the chick's organs, which are approximately divided between the two sexes. I suspect that those who blithely claim that all that is necessary is to look at the vent to see whether or not there is a protrusion, have no idea what they are talking about. Such a protrusion is no guarantee of maleness, and vent sexing of day-olds, in my view, is not a viable practice for the amateur.

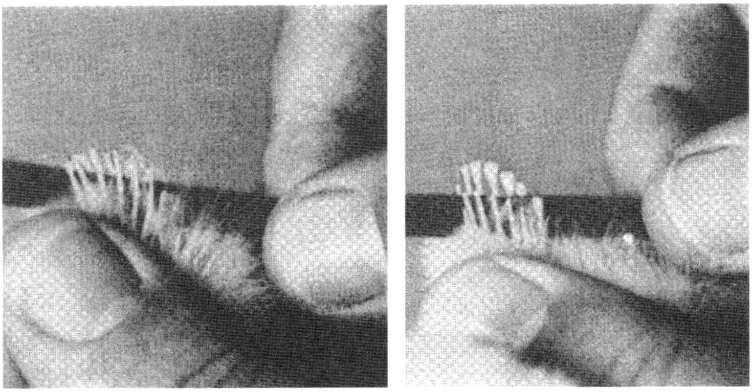

Sexing chicks by comparison of the wing feathers. The primary feathers on the left are the same length or shorter than the covert feathers, indicating a male.

On the right, the primary feathers are distinctly longer than the covert feathers, indicating a female. *Hubbard Poultry.*

Feather sexing

The difference in the growth of primary feathers, particularly in strains of table chickens, is an efficient way of determining sex. Heavier breeds, which include the commercial broilers as well as traditional breeds such as Rhode Island Red, Dorking and Light Sussex, tend to be slower to feather than the lighter egg breeds such as those based on Leghorns. The young show a pattern where the males are slower to feather than the females. Examining the tip of the wing of chicks illustrates this. Cockerels have primary feathers which are the same length, or shorter than the covert feathers, while pullets have distinctly longer primary feathers.

Tail feathering

When birds are a little older, say 3-4 weeks, the more rapid growth of the pullet's tail is a difference to note.

Other methods

Dowsing: This method definitely comes under the category of popular mythology, and I make no claims for it either way, as I have never tried it. However, there are some who swear that they are able to sex day-olds by holding a pendulum (any weight) on the end of a strong thread, and holding it above the head of the chick. If the weight moves in a circle one way, the chick is male, and if it circles in the other direction, it is female. Unfortunately, there is no general agreement about which direction is which, so I leave it to you to experiment.

Leg movements: This is also a popular belief and was related to me by a friend from the next village. If you gently pick up a chick by the head (if you do it carefully it will cause no distress) the females will draw their legs upwards, while the males will kick downwards. Again, I have never tried it.

Head movements: In similar vein, there are some breeders who claim that if you hold a chick upside down, a male will crane its neck up while a female always looks down. Sounds like wishful thinking to me!

Sexing different birds

Ducklings: Vent sexing of ducklings is somewhat easier than that of chicks. The technique is similar and in the males there is a definite penis as distinct from a vague genital eminence.

Goslings: Goslings are similar to ducklings in that the penis of the male is more apparent, but some difficulty may be experienced in handling the gosling, mainly because of its larger size. In all cases, it is essential to have a bright light and to handle the birds with care in case permanent genital damage or prolapse is caused.

Guinea fowl: This heading should read 'not sexing guinea fowl' for, to my knowledge, no-one has yet been able to establish a reliable method of sexing guinea fowl keets. Vent sexing is unreliable and the only course seems to be to wait until adulthood. Then the slightly heavier male with its larger wattles is distinguishable from the lighter female with its finer, more slender head. The call is also different in that the male has a one note call while that of the female is made up of two notes.

Quail: Coturnix laying quail and the Japanese varieties are easy to sex from the age of three weeks onwards because the reddish brown chest feathers of the male become apparent. The male is also smaller, with more definite colour markings than the female. The coloured varieties such as English White, American Range and Tuxedo do not have an apparent difference in feathering and the only way to distinguish them is to watch out for sexual behaviour from the age of six weeks onwards. The male has a more domed vent than the female and also produces a 'foam ball' which looks like shaving cream and is regularly deposited on the ground. Vent sexing is not possible with quail chicks.

Peafowl: I have never bred peafowl so am not able to pass on any personal experience in this respect. However, a friend who does, tells me that she checks the length of the legs. Those of the young males are slightly longer than those of the same aged females. Later on, differences in plumage and behaviour become apparent.

Pheasants: Again, I have never reared pheasants so can only pass on advice I have been given by pheasant rearers. They claim that vent sexing is reasonably reliable, with the penis of the male being distinguishable at an early age.

Ostrich, **Emu** and **Rhea**: With large, expensive birds such as ostriches, emus and rheas, the tendency is increasingly to use DNA sexing techniques. At time of writing these are still relatively expensive, but will inevitably become cheaper as the technique becomes more widespread.

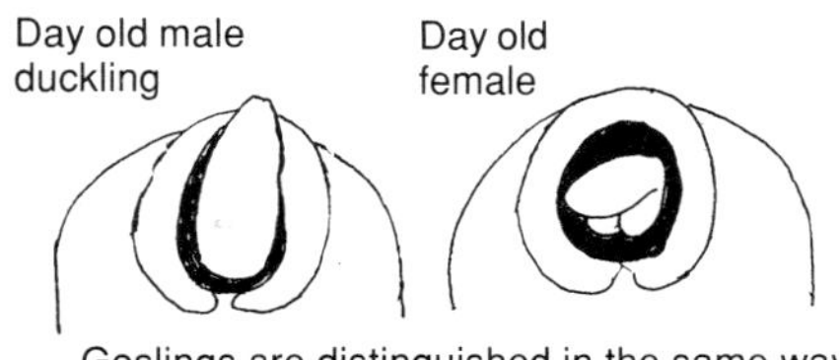

Goslings are distinguished in the same way

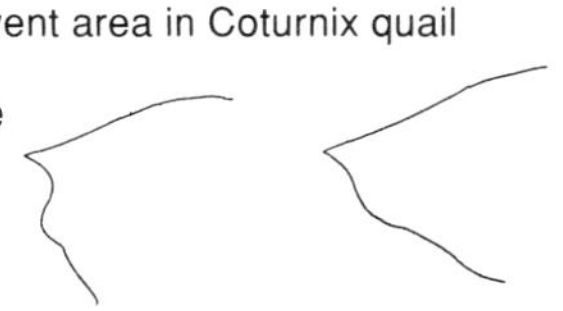
Side view of vent area in Coturnix quail
Left: Male
Right: Female

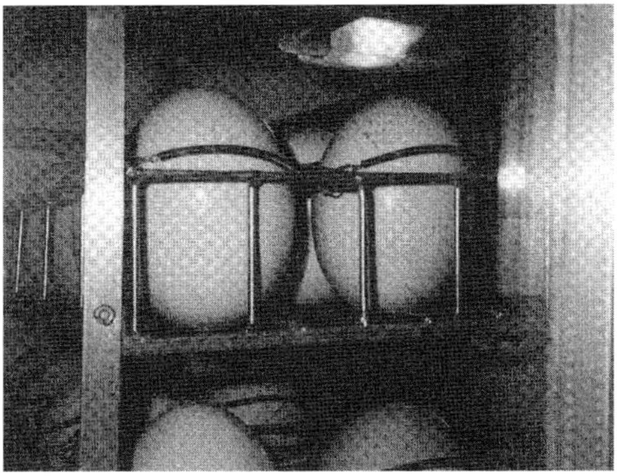

Top left: English White goose with her week old goslings. *Martin Lynch*
Top right: One of the author's Chinese Painted Quail
Middle left: Emu eggs are black and cannot be candled
Middle right: Ostrich eggs half way through the incubation period
Bottom left: Emu breeder
Bottom right: Some of the author's 6 week old turkey poults about to go out on range

Part 3
Different species

"What is that strange bird", said Rupert.

Most of the information in the book is relevant to all birds, although the emphasis is on domestic poultry. It may be useful, however, to be able to refer to a particular section where all the salient points are included for a specific bird.

Ducks

Domestic ducks are not generally reputed to be good mothers, although this varies tremendously, not only with different breeds, but also with individual birds. In my experience, they will incubate their own eggs, particularly in their second year. My Khaki Campbells often used to make nests in the field hedge, emerging with a trail of fluffy ducklings behind them. Broody hens will also incubate duck eggs.

Wild and ornamental ducks will incubate, and rear their own young, as long as the right conditions are available to them. These will vary depending on the type of duck. As a general rule, two types of nest box are necessary, depending on the breeds - those placed on the ground and those above ground level. The ground level ones need to be placed where natural vegetation will conceal and camouflage them. A wide range of containers can be used, from an abandoned milk churn on its side, with turves to support and camouflage it, to purpose-made ones available from poultry and game equipment suppliers. Above-ground nest boxes with entry ramps can also be purchased ready-made, or a home-made one such as a barrel wedged in a tree can be utilised. Above-ground nests are favoured by those breeds which normally nest in holes and tree crevices. Particularly shy breeds may require an artificial island or raft on which to site the nesting box. In this case, stakes can be driven into the pond or lake bottom and weldmesh attached as a flooring. Turves placed on the weldmesh will soon intermesh, providing a natural island base.

If wild or ornamental breed eggs are being incubated artificially, the breeding pairs or trios will need to be moved into breeding pens in late January to February. From here, the eggs can be removed, marked and placed in the incubator. Alternatively, broody hens can be used.

The Pekin duck has rounded eggs which do not have an obvious air sac end. This makes it more difficult to control the weight loss/humidity ratio, and also for the young to pip, according to a waterfowl breeder. It is not unknown for them to pip at the end without an air cell so that they may suffocate unless helped out. The incubation period may also take longer, with 30-32 days being allowed before giving up. For most ducks it is 28 days.

Most incubators are suitable for duck eggs. Store the eggs for no more than a week at a temperature of 15-18°C and a relative humidity of 75% until they are placed in the incubator. For incubation a temperature of 37.5°C is required in the centre of the egg, with a humidity of 58% until pipping. For hatching, the optimum temperature is 37.0°C in the centre of the egg, with an increased humidity of 75%.

Any of the brooders used for chickens are suitable for ducklings, as are the rearing pens or movable arks, but ducks are hardy and should be allowed out as soon as possible. One 250 watt brooding lamp will normally cater for 30-40 ducklings, depending on species. Waterfowl starter ration can be given until the birds are well feathered and grown.

The biggest problem with brooding ducklings is their tendency to splash water everywhere. Suspended drinkers are necessary otherwise they will try and swim in them. Remove areas of damp litter and replace regularly, to avoid health problems.

Muscovy ducks are, of course, renowned for their mothering abilities and I have never needed to incubate their eggs in an incubator. Where they are, the conditions required are slightly different. Following the same storage conditions as the eggs of other birds, those of the Muscovy need an incubator temperature of 37.5°C, with a relative humidity of 60%. At pipping which takes place around day 31, the temperature is reduced to 37.0°C and the humidity increased to 75%.

Geese

Domestic geese are normally extremely reliable in incubating and hatching their own eggs, and the only problem I have ever encountered is that of nest sharing. This can be annoying because the two sets of eggs may become mixed up, and if they are at different stages of incubation may be deserted when the earlier ones hatch. If geese can be persuaded to use nests in a house, so much the better. Alternatively, artificial incubation can be used. Taking away eggs does encourage more.

Incubators which are purpose-made for goose eggs are available. These have deeper shelves to cater for the tall goslings. Temperature requirement in the incubator is 37.5°C until pipping, with a humidity of 55%. From day 27 the temperature is reduced to 37.0 °C in the centre of the egg, with an increased humidity of 75%. Follow the manufacturer's instructions precisely and there should be no problem.

Artificially hatched goslings need protected brooder conditions until they have lost their yellow down and grown their white feathers. One 250 watt brooder lamp will normally do for 25 goslings. They are much hardier than chickens, however, and the sooner they can go out on range the better. They need exercise to avoid leg problems and they are natural grass grazers anyway.

A waterfowl starter ration is given initially, with clean, fresh water (for drinking only, not swimming). It is important to avoid wide drinkers otherwise the goslings will try and go for a swim.They quickly learn to peck at anything resembling grass (including spiky, damp down feathers), so their access to fresh, clean, short-growing pasture is important. Keep an eye open for gizzard worm infestation if the pasture has been used before. If a young bird on pasture begins to be listless, sits down a lot and is obviously off colour, suspect gizzard worm and get veterinary advice. The only product licensed as a poultry vermifuge in the UK, is *Flubenvet* which is added to the feed. One the whole, however, geese require far less medication than chickens, and chick crumbs with added coccidiostat which are normally given to chickens may be detrimental to them.

Turkeys

Since the rapid development of the turkey industry, many modern hybrid strains of white turkeys are now so huge that natural mating is impossible; artificial insemination is the norm. Yet, the smaller or more traditional turkeys will readily sit and brood their eggs. On a visit to the Dordogne area of France, I came across a farm where a beautifully glossy black turkey hen was taking her brood for a walk, quite unconcerned by the excited gaze of a visitor from Britain.

Second year hens are more reliable than first year ones, and become broody very easily once they have laid a clutch of eggs which has been left undisturbed. A hen turkey will lay more if her eggs are removed daily. For this reason, many people prefer to use broody hens to hatch the eggs, thus allowing the turkey to produce more.

As far as artificial incubation is concerned, most incubators are suitable. Again, the manufacturer's instructions are crucial. The optimum incubation temperature in the centre of the egg is 37.5^{0}C, with a humidity of 55%. At pipping which is around day 25, the temperature is reduced to 37.0^{0}C in the centre of the egg, with an increased humidity of 75%.

Brooding conditions are similar to those provided for chickens, with a brooder lamp and containment ring. A thick layer of wood shavings provides good flooring, but ensure that all the poults have easy access to, and are eating the feed, otherwise they may peck the litter and become impacted. Insoluble grit can be provided to help deal with impaction.

A turkey starter ration is normally given on an ad-lib basis. This contains an additive to prevent the incidence of Blackhead, *Histomoniasis*, a protozoan infection. Additive-free rations are also available from specialist feed suppliers, but if this is fed, every effort should be made to ensure that turkeys do not have access to ground used by other poultry.

Guinea fowl

Guinea fowl are similar to turkeys in their management requirements, and are also at risk from Blackhead. Incubation is usually artificial, although second year hens will incubate and brood their own young if given the opportunity to do so. A nest box such as that provided for a large broody hen is suitable.

The optimum temperature for the first part of incubation is 37.5°C in the centre of the egg, reducing to 37.0°C at pipping time on day 25. Humidity level is 55% up to pipping, rising to 75% while the keets are emerging.

Brooding conditions similar to those provided for other chicks are suitable, but the keets should be allowed outside as soon as weather conditions permit, returning to their protected and heated conditions on demand.

A game fowl starter ration is a suitable feed, along with clean, fresh water. An important part of their diet under normal conditions is insectivorous, so the starter ration needs to be fairly high in protein.

Guinea fowl are rather flighty birds and care must be taken that the young keets do not panic and rush to one area of the brooding area. It is not uncommon for birds to be smothered by piling up in this way.

Quail

Most general purpose incubators will take quails' eggs and I have successfully hatched them in a DIY incubator as well as in a bought model. I have not used one with an automatic turning facility, but a commercial quail breeder told me that his hatching rate had shown a considerable increase since he bought such a machine. It was his view that quails are particularly vulnerable to erratic development as a result of inadequate turning of the eggs.

Aviary housed quails will often incubate their own eggs, particularly if natural, sheltered conditions are available to them. Most quails, however, are incubated artificially. The incubation period varies according to the breed (See the Table on page 49), but the conditions required are identical.

The optimum temperature at the centre of the egg is 37.5°C, decreasing to 37.0°C at pipping. Optimum humidity is 45% for the main incubation period, increasing to 75% at hatching time. The patterned nature of quails' eggs makes them difficult to candle effectively.

Once hatched, the young can be fed on chick crumbs and fresh water, and given normal brooder conditions. A word of warning here! Young quail are tiny and great escapees. They also drown easily in normal water containers, so it is important to put stones in the drinker, to reduce the depth and so, the risk. Some ideas for brooders suitable for tiny occupants are given on page 62.

Pheasants

Pheasant eggs can be reared under broody hens but most are artificially incubated these days, and the practice is to use separate incubators and hatcher. Breeding stock of game fowl are caught early in the year and kept in pens until a reasonable outlay of eggs is achieved. This will be approximately 30 eggs per hen bird from a breeding ratio of 1 cock to 6 hens in a breeding pen measuring 20' x 20'. This will need low perches and a shelter at one end. Coops sheltered by cut branches also provide shelter and privacy, to encourage the hens to lay.

A pheasant breeder ration should be given ad-lib in hoppers protected from the weather, with fresh water available at all times. Incubate the eggs as quickly as possible in a cabinet setter with later transference to a hatcher. They are in the former for 21 days and moved to the hatcher for the last 3-4 days.

Incubation temperature needs to be 37.5^{0}C in the centre of the egg, with a humidity of 50%. At pipping on day 21, the temperature is very slightly reduced to 37.0^{0}C at the centre of the egg, while humidity is increased to 75%.

Once hatched and dried, normal brooder conditions should be provided, with a pheasant starter ration being given for the first few weeks, gradually going over to a rearing ration. According to a pheasant breeder, the average hatching rate for pheasants in these conditions is around 70%.

Partridges

Partridges require conditions similar to those for pheasants, although it has been claimed that the eggs will stand storing for a longer period (up to two weeks) before incubation takes place. They still require careful storage conditions and placing the egg carton in a perforated plastic bag may help to keep the eggs viable for this longer period.

Incubation temperatures are 37.5^{0}C and 37.0^{0}C respectively for the setting and hatching stages, while the humidity requirement is 47% up to pipping stage, followed by 75% for hatching. Brooding requirements are as for any chicks, and a pheasant starter ration will give them a good start.

Peafowl

A peahen will lay a clutch of 7-10 eggs and then sit on them unless the eggs are removed every day. In this case, she will lay up to around 32-48 eggs in a season.

The eggs can be hatched under broody hens, given to the peahen herself, or incubated artificially. The best situation for a broody peahen is a nest within a house or shelter where she can have as much privacy as possible. Providing a broody box does not work for it is usually too confined, although it is possible to erect a partition in the corner of a shed. Here, wood shavings can be given as a nest liner, and provided she has peace and quiet, should prove to be a good broody.

With artificial incubation, the eggs require a temperature of 37.5^{0}C for the first half, reducing to 37.0^{0}C, while humidity up to pipping is 55%, increasing to 75% for hatching. Normal brooding conditions are required by the young, including a high 28-30% protein ration. A pheasant starter or turkey starter ration is normally suitable. This can be reduced to 26% protein rearing ration after five weeks.

Ostriches

Ratites need large exercise and ranging pens, and the breeding birds will perform better without lots of distraction. Sheltered, private areas are therefore an asset. The females also need a soft, sheltered place in which to lay: a barn with sand or wood shavings in a dark corner is ideal. Removing the eggs encourages the laying of more. As soon as possible, the eggs should be candled so that the position of the air cell can be established in the rounded eggs. They are then stored, air cell upwards until incubation which should be as soon as possible.

Incubation temperature is 36.0^{0}C, while humidity is 30%. The large eggs need to lose up to 15% of their weight, and this can be a problem in areas such as Britain where the natural humidity is often high. The ostrich is a desert bird and the initial low humidity is essential for the thick-shelled egg to lose sufficient moisture to develop an adequate air sac. Dehumidifiers are normally used in the incubators. Regular weighing and candling of the eggs are essential in order to monitor development..

At pipping, the temperature is reduced to 35.5^{0}C, while humidity is increased to 75%. Regular weighing and candling of the eggs is advisable. The total incubation period is 42 days, with pipping occurring after 35 days.

Once hatched, the young require protected conditions and should be fed on a special ostrich starter ration which is available from specialist suppliers. They have a tendency to eat foreign objects which can lead to impaction. As they are prone to leg weakness, it is important to give them non-slip flooring such as ribbed rubber mats, and to allow as much exercise as possible. The use of these mats also means that they do not have litter to eat. Going outside as early as possible is advisable, as long as weather conditions permit, and that they are provided with windbreaks and shade. The exercise is essential for leg development and hardy growth, and they have the protected conditions available to go back to in the evening.

Emus

At two years of age, emus will be ready to breed and they mate as breeding pairs, although kept in a communal flock.The female is dominant and it is she who selects a mate and defends the teritory. The male sits on the nest to incubate the eggs.

Emu eggs can be incubated artificially. They are produced in the winter months, between November and March, and there are 30-50 produced per season. The shells are almost black and cannot be candled. As the eggs are equally rounded at both ends, the position of the air cell cannot easily be determined. One way of doing so is to tap the egg and listen for the different sound produced in the air cell area. Until the air cell is identified, the eggs should be stored on their side. Once the position of the air cell is identified, however, the eggs can be stored with the air cell at the top.

A temperature of 36.0^{0}C at the centre of the egg is required for incubation, with a relative humidity of 40%. At the pipping stage which is at day 46, the temperature is reduced to 35.5^{0}C and the humidity increased to 75%. Hatching normally takes place on day 50, although 52 days is not unknown.

Around 15% weight loss is required during the incubation period, and this needs to be established by regular weighing.

Normal brooding conditions are required, with heat being made available until the chicks are three months old, although they should be allowed to exercise outside as soon as possible. The floor of the brooding area should be non-slip and ribbed rubber mats have been found to provide good traction to prevent slipping.

Impaction, or the tendency to eat foreign objects that cause digestive blockage, is not as marked as it is with ostrich chicks, but care should be taken to avoid this. Purpose-made emu rations are available and a starter mix should be made available. Chopped greens can also be given with free access to insoluble grit.

Rheas

Rhea eggs are lemon yellow in colour, and require an incubation temperature of 36.5^{0}C at the centre, reducing to 36.0^{0}C for harching. A relative humidity of 35% is required until the pipping stage at 33 days, when it is increased to 75%. 15% of the egg weight needs to be lost during incubation, to ensure proper air cell development. Hatching normally takes place on day 36, but may be a little earlier or later.

The young require normal brooding conditions with a heat lamp, but should be encouraged to exercise as much as possible. Ribbed rubber mats for the floor, to prevent slipping and leg damage as for ostriches and emus, are advised .

Rheas can be fed an ostrich ration, and alfalfa pellets if forage is not available. Impaction can be problem, so care is necessary to prevent access to foreign objects.

Parrots

Female parrots only have one ovary, on the left side of the body. Even so, they can be encouraged to lay a considerable number of eggs by removing them as they are laid, and stored prior to incubation. When first laid they are soft, but quickly harden.

Depending on the species, the incubation period is 18-29 days and the eggs need to lose up to 15% of their weight to ensure adequate development of the air cell. Incubation temperature is 37.5^{0}C at the centre of the egg, with a relative humidity of 50%. At pipping, which is around 3 days before hatching (15-26 days depending on species), the temperature is reduced to 36.6^{0}C and the humidity raised to 75%.

Parrot chicks are quite helpless at hatching, with eyes closed and often naked. They are extremely prone to chilling and great care must be exercised in brooding. On page 63 there is a photograph of a parrot chick being brooded in an incubator which has been adapted as a brooder by the addition of a higher lid. Starter and growing rations for parrots are available from specialist suppliers.

Part 4
What went wrong?

The reasons why eggs do not hatch successfully are many and varied. They include:

Congenital defects. These appear if the cock and hen are too closely related and in-bred. Suspect that this is the case if several of the chicks are deformed.

Some birds may carry *lethal genes*. These are normally recessive in that they do not appear unless both parent birds are carrying the lethal gene. Some examples are crooked neck, dwarf limbs, lack of wings.

Infertility. This is usually because one or both of the parents are old, ill, over-fat or over-bred. The latter is all too common amongst some of the show strains where too much emphasis has been placed on breeding for appearance rather than vigour. Infertile eggs can be detected at 7 days into incubation by candling. They can then be removed from the incubator so that they do not provide a source of infection.

Inherited diseases. These may not affect the parents, but could well be lethal to the chicks. Anyone who is in the business of selling birds or fertile eggs, really should have the breeders blood-tested by a vet to make sure they are free of inheritable diseases such as Fowl Typhoid that can be transmitted via the egg. They should also be tested to make sure they are free of Salmonella.

Vitamin and mineral deficiencies in the parents' diet can cause deficiency conditions in the chicks. (See later).

Rough handling. Reference has already been made to the necessity of treating the eggs gently at all stages of handling.

Inadequate hygiene. Not observing hygienic procedures, such as dipping eggs in sanitant before incubation and ensuring that the incubator is clean, will greatly increase the chances of infection.

Poor incubation practice. Again, reference was made earlier in the book to the importance of siting and setting up the incubator properly, testing it before use, checking temperature, ventilation and humidity, regular egg turning, and candling or weighing.

Poor brooder practice. A much higher success rate will be had by providing proper protected conditions, sufficient warmth and ventilation, adequate food and water, and vaccination protection of the chicks against important diseases. See page 66.

Some of the specific problems that may be encountered are:

Clear eggs with no development

• *The eggs were infertile* - the male or female (or both) are infertile, ill, too old, or inadequately nourished (they could also be over-nourished and too fat). The male may be suffering from environmental stress, have too many females or have too much competition leading to fighting. Frost damage to the comb can affect the male. Provide adequate accommodation, conditions and diet. Use good stock and observe mating ratios.

• *The eggs were stored too long before incubation.* Ideally incubate no later than 7 days after lay. Store at 15-18°C and 75% relative humidity. If to be kept longer, reduce temperature to 12-15°C. If much longer storage is unavoidable, place eggs in polythene and lower temperature to 10°C.

• *The eggs were damaged.* Gather frequently and handle gently.

Partial incubation with blood ring

• *Bacterial infection* such as *Escherichia coli*, where the egg has been subject to damage, cracks, dirt. Handle properly and follow hygienic procedures of incubator disinfection and egg dipping with sanitant.

• *Viral infection.* Use eggs from disease-free sources. Do not use breeders that have not been blood-tested and vaccinated.

• *Improper storage.* See details above.

• *Too high or too low a temperature* setting in the incubator. The optimum temperature is 37.5°C at the centre of the egg (36°C for ratites). Check accuracy of thermometer and thermostat and follow the manufacturer's instructions precisely.

Bad eggs

• *Infertile.* Candle at 7 days and remove 'clears'.

• *Bacterial infection.* Damage to the egg shell or inadequate hygiene within the incubator can allow the entry of *Escherichia coli* bacteria. It is often accompanied by a horrible smell.

Large, soft-bodied chicks

• *Mushy chick disease.* Also called Yolk sac or Navel infection (Omphalitis). This is caused by the entry of *Escherichia coli* bacteria. Death usually occurs within the shell, but it can be shortly after hatching. The abdomen is often distended because the yolk sac is not fully absorbed. The navel area may be discoloured and in severe cases there is an offensive odour. There is nothing that can be done, but it could have been prevented by correct egg handling, incubator disinfection and egg dipping in a sanitant.

Dead embryos positioned to one side

• *Eggs not turned.* Turn at least 5 times a day. Ensure that eggs have enough room and are properly positioned, either on their side or with blunt end upwards.

Early death of embryos

- *Temperature too high or too low*. Check thermometer and thermostat.
- *Inadequate ventilation*. Provide proper aeration and check before starting. If at a high altitude increase aeration and temperature slightly.
- *Irregular turning*. Turn the eggs at least 5 times a day if the incubator is a manually turned one.
- *Viral infection*. Use eggs from disease-free sources. Blood-test parent birds.
- *Inherited low hatchability*. Use good breeders.

Chicks hatch early (often with blood on the navel).

- *Temperature too high*. Check thermometer and thermostat. Reduce temperature to 37°C for the pipping and hatching (35.5°C for ratites). Increase ventilation.

Slightly rough, unhealed navel

- *Temperature too high*. As above
- *Early stages of Navel infection*. Treat chicks with disinfectant on the navels and put in a separate brooder from the rest of the hatch. They may survive. Avoid future problems by using proper handling and hygiene procedures.

Chicks fully formed but dead in shell

- *Deformity of beak*. Avoid in-breeding parent birds. Ensure that they have sufficient calcium and other nutrients in the diet.
- *Incorrect humidity*. Check level for the appropriate species and stage of incubation in the Table on page 49.
- *Incorrect temperature*. As above.
- *Inadequate ventilation*. Check fan or open ventilation holes more, as appropriate.

Late hatching

- *Temperature too low*. Maintain temperatures as recommended above.
- *Cold patches in incubator*. Use a fan-assisted one next time or check temperatures more carefully in all areas of the incubator, and adjust ventilation holes and insulation accordingly.

Embryos smeared with cell contents

- *Hereditary disease*. Use different breeders.
- *Incorrect humidity*. Check as above.
- *Bacterial infection*. Follow proper handling and egg sanitation procedures.
- *Inadequate ventilation*. As above.

Pipped chick unable to emerge

- *Insufficient humidity*. Dampen slightly to stop chick sticking to dried membranes.

Small chicks

- *Small eggs*. Use medium-sized eggs for incubation.
- *Temperature too high*. Adjust correctly for incubation and hatching stages for specific eggs (See Table on page 49).
- *Small or poorly fed parents*. Select breeders appropriately and feed a proper breeder's ration with optimum nutrients.

Grossly deformed chicks

- *In-breeding*. Select different parents.

Clubbed end to down feathers

- *Vitamin B_2 deficiency in parents*. Feed proprietary breeder's ration. Allow parents to grass range and if making up your own ration, feed a supplement of chopped boiled egg, Marmite, yeast or ground-up yeast tablet.

Curly toes

- *Lack of Vitamin B_2 (Riboflavin)*. Feed parents as above.

Splayed legs

- *Rickets*. The affected chick has rubbery legs and is unable to stand without difficulty. The breast bone is often twisted. The beak may also be pliable so that death in the shell may have resulted because it could not peck its way out. It is caused by a shortage of vitamin D in the parents' diet leading to inability to metabolize calcium and phosphorus. Feed breeder's ration. Allow parents to grass-range in the sunshine and give crushed oystershell and cod liver oil supplements.
- *Slipping on unsuitable surface*. Use the correct tray insert for the particular type of egg in the incubator. In the brooder, avoid slippery surfaces such as smooth newspaper. Use a thick layer of wood shavings. Ratites are recommended to have ribbed rubber mats and exercise as soon as possible. Ducks and geese also need plenty of space and exercise.

Splayed legs can be corrected by the use of splints, as long as the damage is not too severe. In the case of small chicks, match sticks work well. Tape them on, making sure that the wrapping is not too tight. After a couple of days, the problem is usually rectified. Severe cases should be put down quickly and humanely.

Inability to coordinate movements

- *Crazy chick disease*. This is more correctly named *Encephalamasia* and is caused by insufficient vitamin E in the parents' diet. The chicks can only look up and not down. Feed parents a proper breeder's ration or give wheat germ extract.

Very bright yellow down

- *Excessive fumigation*. This is not a problem that small breeders are likely to encounter because fumigation with formaldehyde is not a small scale procedure.

Generally low hatches

• *Low fertility.* See previous comments about infertility.

• *Too little protein amino acids in the parents' diet.* A properly balanced breeder ration provides all the necessary nutrients, or give supplements of hard-boiled eggs.

• *Vitamin B_{12} (cyanocabalmin) deficiency in the parents' diet.* This leads to inadequate blood formation and often shows up where guinea fowl, for example, are not allowed to range sufficiently to catch insectivorous prey, and are not fed a breeder's ration to compensate. Confined ducks without access to a pond where they can catch insects may also have the deficiency, unless they are receiving proprietary rations. Comfrey leaves are the only plant source of vitamin B_{12} and can be fed to a wide range of birds.

• *Lack of trace elements in parents' diet.* Give breeder's ration or ground-up multi-vitamin tablet as a supplement.

Chicks gasp for breath, opening and closing their beaks

• *Brooder pneumonia (Aspergillus fumigatus).* This is caused by fungus spores which can so easily be introduced via rough handling, lack of hygiene procedures and hairline cracks in the egg. Make sure that the parent's nesting material is clean and dry and that eggs are dipped in sanitant.

Affected chicks often try and drink more than normal. It can cause high mortality in the first few days of life and there is nothing that can be done. Again, it could have been avoided.

• *There are many other respiratory infections* which may cause wheezing and shaking of the head. Try to ensure that the chicks have warm, protected conditions and no draughts, yet have sufficient ventilation. Use only blood-tested breeders.

Trembling chicks

• *Epidemic tremor (Avian encephalomyelitis).* This is a virus infection transmitted to the egg from infected breeders. It normally shows up in chicks a few days after hatching when they sit back on their hocks and visibly tremble. The mortality rate is high. Breeding birds can be vaccinated against the condition.

Pasted rear ends

• *Pullorum disease (Salmonella pullorum or white diarrhoea disease).* This can be passed on to the chick from an infected hen, via the egg. Again it shows the importance of using parent birds that have been tested.

The chick has a pasted rear end, with the typical white bacillary diarrhoea. Mortality is high and infected chicks can, in turn, infect others in the brooder.

Inability to eat and drink

• *Lack of example.* Artificially hatched birds which do not have the example of the mother bird may be slow to eat and drink. Place feed in shallow containers to make it easy to find, and drop some on a sheet of paper to imitate the action of the mother. Dip the chick's beak in water to teach it to drink.

• *Impaction*. This is where litter has been eaten causing a blockage. Dose the chick with *pharmaceutical* grade liquid paraffin and make grit available. Place paper towels over the brooder litter until they have learnt to feed and drink, then remove.

Pecking

• *Wet feathers*. Arrange drinkers so that young birds cannot get into them or splash water in the feathers. The 'spikiness' of wet feathers invite pecking from other birds.

• *Blood or other marks on feathers or toes*. This can also act as an invitation. Treat the chick immediately with disinfectant to give a repelling taste, as well as healing.

• *Insufficient space*. Ensure that young birds are not over-crowded.

• *Insufficient food and water*. These should be available on an ad-lib basis.

• *Light is too bright*. Newly hatched chicks should have light 24 hours a day for the first few days, but it should not be too bright. Dim the lights slightly if there is a problem (as long as they have all learnt to feed and drink). Thereafter, gradually reduce lighting to normal daylight hours, or to 8 hours if the birds are to be layers. At point of lay increase light to 15 hours (natural + artificial) to induce laying).

• *Boredom*. Let them range in protected runs as soon as possible and provide suspended greens to peck at.

• *Parasites*. Make sure there are no mites or other external parasites present. Ruffle the feathers over a sheet of white paper. Are there small black specks moving about? If young birds are being brooded by a broody hen, was she treated before sitting?

Droopiness, ruffled feathers and blood in droppings

• *Coccidiosis*. This is caused by a protozoan parasite. It is often associated with wet litter. Give chick crumbs which contains a coccidiostat to give them a good start in the vulnerable period. (Some ducks and geese may react to the coccidiostat and it is better to give them a starter ration specifically for waterfowl. Ratites and game birds also have their own specially formulated starter rations).

Suffocated chicks

• *Piling*. This is where birds pile into a corner as a result of panic or an attempt to keep warm, and some are smothered. Place enough lamps in central positions. In large floor areas erect close-meshed netting across the corners. Always enter the brooding area slowly and talk quietly to the chicks.They learn to recognise their owner's voice. Similarly, the breeder can distinguish between contented cheeps and distressed sounds.

Finally, it's important that young birds without a mother are reared with their own kind so that the correct behaviour is imprinted. A solitary bird will treat the owner as its mother and may be incapable of mating with its own kind because the appropriate imprint is lacking. Geese are particularly vulnerable in this respect.

Reference Section

Many of the following incubator and hatcher suppliers also sell associated equipment such as brooders, lamps, brooding rings, dehumidifiers, candlers, thermometers, hygrometers, etc.

Incubator Suppliers
A B Incubators Ltd, Tel: 01473 780050, www.abincubators.co.uk
Brinsea Products Ltd, Tel: 08452 260120, www.brinsea.co.uk
Bristol Incubators Ltd, Tel: 01454 228730, www.patrickpinker.com
Curfew Incubators, Tel: 01621 741923, www.cyclex-uk.co.uk
Grumbach Incubators, Tel: 01246 264620, www.alphahatch.com
Mardle Products Ltd, 01493 249140, www.mardleproducts.co.uk
M S Broedmachines, 07796 267666, www.msincubators.com

USA
Humidaire Incubators, Tel: +1 937 996 3001
Kuhl Corporation, Tel: +1 908 782 5696, www.kuhlcorp.com
Lyon Electric Company, Tel: +1 619 216 3400, www.lyonelectric.com

Appendix

Egg capacities

Some incubators describe their capacities in chicken egg sizes. If you want to know how many eggs of another bird will fit, the following table gives you the approximate equivalents.

Bear in mind that automatically turned models will normally have less egg room than manually turned ones.

Eggs	*Equivalent*
Chicken	60
Duck	45
Goose	24
Turkey	40
Pheasant	90
Quail	150

Temperature conversion

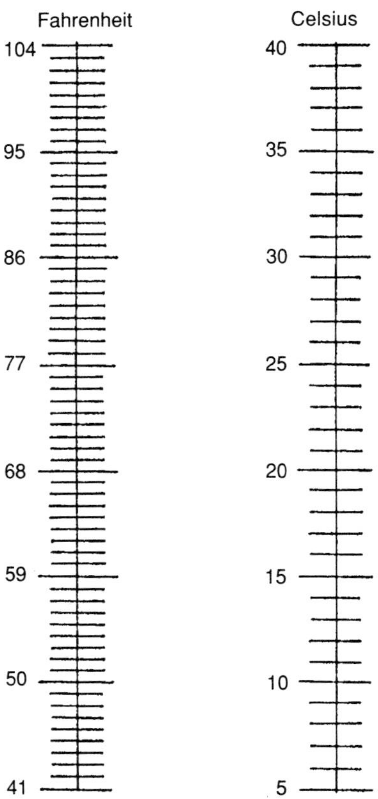

Mating ratios

The following is a general indication of the mating ratios of females to males in breeding flocks. Bear in mind that these are generalisations only, for ratios may differ according to confinement and general flock management.

Birds		*Ratio of females to males*
Chickens	Light	10:1
	Heavy	8:1
	Bantams	12:1
Ducks	Light	10:1
	Heavy	7:1
Geese	Light	5:1
	Heavy	3:1
Turkeys		6:1
Guinea fowl		8:1
Quail		6:1
Pheasants		6:1
Partridges		6:1
Peafowl		5:1
Ostrich		3:1
Emu		1:1
Rhea		3:1

To convert from Fahrenheit into Celsius:

Subtract 32
Multiply by 5
Divide by 9

To convert from Celsius to Fahrenheit:

Multiply by 9
Divide by 5
Add 32

To calculate fertility percentage

Number of eggs in incubator = a
Number of eggs fertile at candling = b

$$\% \text{ of fertile eggs} = \frac{b \times 100}{a}$$

To calculate hatch percentage

Number of fertile eggs = b
Number of eggs hatched = c

$$\% \text{ of fertile eggs hatched} = \frac{c \times 100}{b}$$

(Answer to question on page 2: There are 13 chicks).

Index